Second Edition

# GENERAL CHEMISTRY

## LAB II

CHE 112L

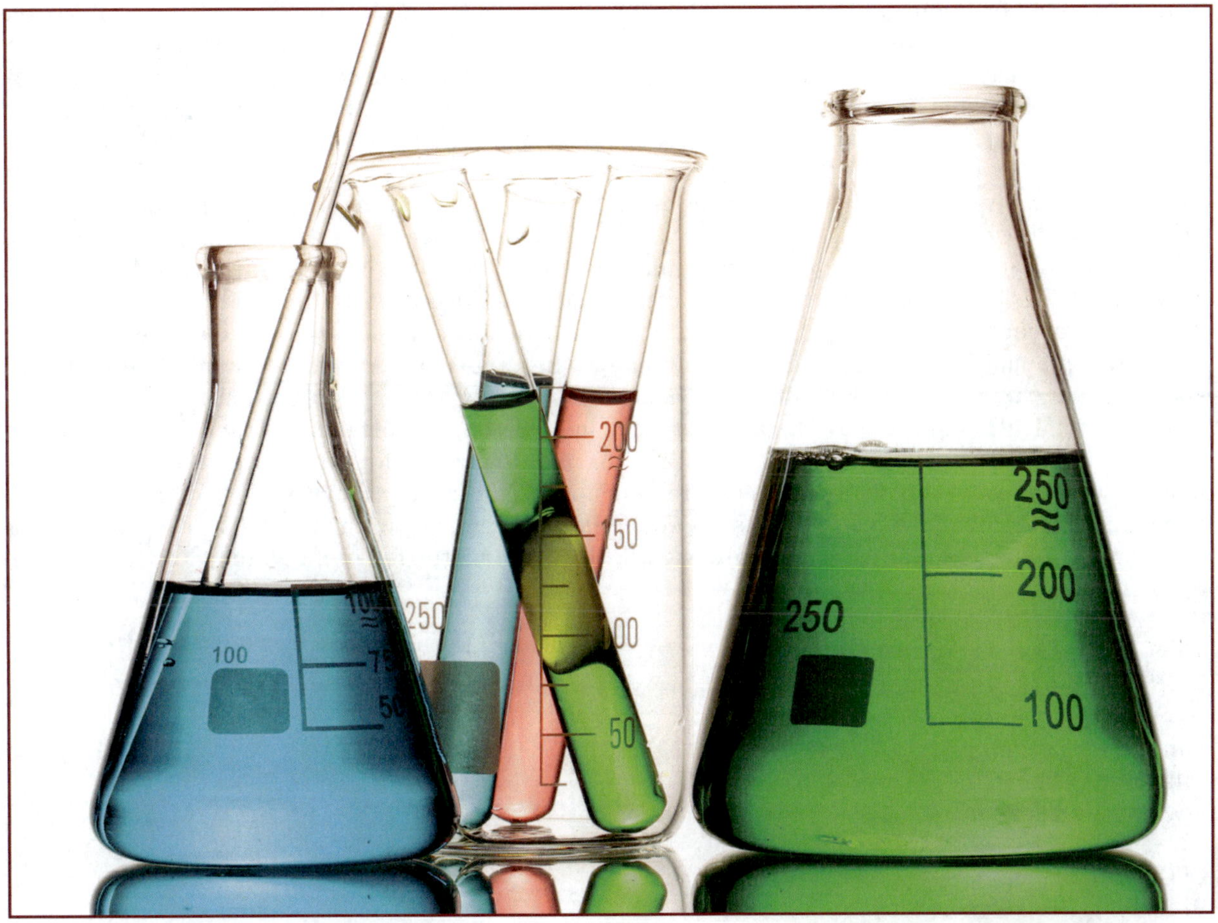


JEROME MAY
DEPARTMENT OF CHEMISTRY
EASTERN KENTUCKY UNIVERSITY

VAN-GRINER LEARNING

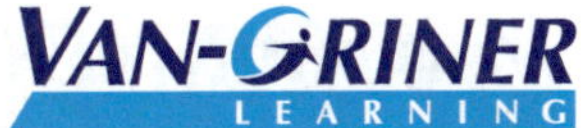

# *General Chemistry Lab II*

CHE 112L
Jerome May | Second Edition
Department of Chemistry
Eastern Kentucky University



Printed in the United States of America
10  9  8  7  6  5
ISBN: 978-1-61740-681-2

Van-Griner Learning
Cincinnati, Ohio
www.van-griner.com

CEO: Mike Griner
President: Dreis Van Landuyt
Project Manager: Chessey Brodbeck
Customer Care Lead: Julie Reichert

May 681-2 W19
310628-321380
Copyright © 2020

# Table of Contents

# Molecular Geometry



## CHE 112L

- An adaptation and extension of the CHE 111L Lewis Electron Dot Diagram (LEDD) laboratory exercise

# 20-QUESTION PRE-LAB ASSIGNMENT

CRN:_________________________________________     Your Name: ___________________________________

Date Submitted: ______________________________     TA's Name: ___________________________________

*Please write your answers legibly in the space provided. Some answers will actually consist of three or four pieces of information that are clearly written on the board during the conversation **or** clearly stated verbally during the presentation. You must include all aspects of the answer to receive full credit. These answers are due at the beginning of the lab meeting.*

**1.** What is the **"second task"** for us to do this week once we have the valid Lewis structures?

**2.** What does the abbreviation LEDDs stand for?

**3.** What do the dots in a Lewis dot structure represent?

**4.** What does the Roman numeral at the top of a column (family) tell us?

**5.** Fill in the blanks: The noble gases do not have any ________________ electrons. All of their electrons are ________________ electrons.

**6.** Draw the LEDDs for ammonia (on the left) and methane (on the right) in the space below.

**7.** What are the three criteria that ***both*** of these LEDDs satisfy?

**8.** What is an **upe?** And how do upe's help us predict the expected valence?

**9.** Which atom(s) do we typically expect to satisfy the duet rule and the octet rule?

**10.** What exactly do we look at when we are ready to select a **geometry?**

**11.** What is an **electron set?**

**12.** How many electron sets are on N in ammonia?

How many electron sets are on C in methane?

**13.** How many of the electron sets are **bonding sets** on N?

How many of the electron sets are **non-bonding sets** on N?

How many of the electron sets are **bonding sets** on C?

How many of the electron sets are **non-bonding sets** on C?

**14.** Fill in the blank: **Four electron sets =** _________________ **electron set geometry.**

**15.** Draw the 3D electron set geometry around N (in ammonia) and around C (in methane) on this 2D piece of paper (just as it was drawn on the 2D classroom chalkboard).

**16.** What do the wedges and the dashes (i.e., the hatured lines) represent in the 2D depiction of a 3D structure?

**17.** What is the molecular shape regulated by?

**18.** What is the electron set distribution about the central N atom in ammonia, and what is the molecular shape of the ammonia molecule?

**19.** What is the electron set distribution about the central C atom in methane, and what is the molecular shape of the methane molecule?

**20.** Oops! There is a misspelled word next to the methane molecule. Identify the misspelled word ***and*** then correct it!

# BACKGROUND

## Drawing LEDDs of Molecules

These steps describe a method that can be used to write valid structural formulas for a majority of molecular formulas.

1. Look at the molecular formula, and decide where to place the atoms in order to draw the LEDD.

   a. The first element listed in the molecular formula will **usually** be placed at the center of the LEDD. (Notable exceptions include compounds with H listed first because, based on valence considerations, H cannot typically be a central atom.)

   b. Elements that there are more than one of in the molecular formula are **rarely** in the center of the LEDD. (Notable exceptions include compounds containing carbon because carbon atoms have the ability to *catenate*.)

   c. The remainder of the atoms should be symmetrically arranged around the periphery of the central atom.

2. Count up all the **valence** electrons each atom brings into the LEDD. (Recall that the Roman numeral at the top of columns IA–VIIA tells us the number of valence electrons for elements in that family.)

3. Begin by placing one **pair** of electrons between the central atom and each peripheral atom.

4. Make sure that each element has its valence requirements satisfied.

5. Check to see that column IVA–VIIA elements satisfy the octet rule and that hydrogen obeys the duet rule.

6. The number of electrons in the final LEDD must equal the total number of valence electrons contributed by all the atoms in the molecular formula.

7. For polyatomic cations, remove one electron per positive charge from the total; for polyatomic anions, add one electron per negative charge to the total.

8. If the octet rule is satisfied, BUT the electron count in the LEDD does not equal the total number of electrons available, then **multiple bonds** will be required in the LEDD.

   1 electron pair = a *single* bond

   2 electron pairs = a *double* bond

   3 electron pairs = a *triple* bond

9. Some column IVA–VIIA elements in Period 3 and below may occasionally exceed the octet rule. (These elements have low-lying d-orbitals which are capable of accommodating the extra electron density.)

10. They may also exceed their expected valence. (Period 2 elements engaged in dative— a.k.a., coordinate covalent—bonding may also exceed their expected valence, but they may *never* exceed the octet!)

11. For any element/atom that **ever** exceeds its expected valence, the actual valence may **never** exceed the actual number of valence electrons that the atom started out with.

12. For CHE 111/112 purposes, the number of valence electrons that should be assigned to transition metals is also the Roman numeral at the top of its column.

13. For CHE 111/112 purposes, the number of "valence electrons" that should be assigned to the noble gases (Kr, Xe) is also the Roman numeral at the top of its column (VIII).

14. The Formal Charge Model can be used to help explain why some atoms violate their expected valence.

15. The sum of the formal charges of all the atoms present must equal the overall charge of the chemical species.

16. The sum of the oxidation numbers of all the atoms present must equal the overall charge of the chemical species.

17. Resonance structures may be necessary to accurately represent empirical evidence like bond angle, bond length, and bond strength.

## TABLE 1  The Shapes of Molecules

| Atomic Orbital Hybridization Name (number of groups) | sp-hybridized (2) | sp²-hybridized (3) | | sp³-hybridized (4) | | |
|---|---|---|---|---|---|---|
| **Molecular Shape** (class) | Linear (AX$_2$) | Trigonal Planar (AX$_3$) | V-Shaped or Bent (AX$_2$E) | Tetrahedral (AX$_4$) | Trigonal Pyramidal (AX$_3$E) | V-Shaped or Bent (AX$_2$E$_2$) |
| **Number of Bonding Groups** | 2 | 3 | 2 | 4 | 3 | 2 |
| **Number of Lone Pairs** | 0 | 0 | 1 | 0 | 1 | 2 |
| **Bond Angle** | 180° | 120° | <120° | 109.5° | <109.5° | <109.5° |

| Atomic Orbital Hybridization Name (number of groups) | sp³d-hybridized (5) | | | | sp³d²-hybridized (6) | | |
|---|---|---|---|---|---|---|---|
| **Molecular Shape** (class) | Trigonal Bypyramidal (AX$_5$) | Seesaw (AX$_4$E) | T-Shaped (AX$_3$E$_2$) | Linear (AX$_2$E$_3$) | Octahedral (AX$_6$) | Square Pyramidal (AX$_5$E) | Square Planar (AX$_4$E$_2$) |
| **Number of Bonding Groups** | 5 | 4 | 3 | 2 | 6 | 5 | 4 |
| **Number of Lone Pairs** | 0 | 1 | 2 | 3 | 0 | 1 | 2 |
| **Bond Angle** | 90° (ax) 120° (eq) | <90° (ax) <120° (eq) | <90° (ax) | 180° | 90° | <90° | 90° |

# DATA SHEET

Name: _______________________________  Grade: _______________________________

Date Experiment Performed: _______________________  Days Late: _______________________________

CRN of Lab Section: _______________________________  Instructor's Initials: _______________________________

| TABLE 2   The 14 Molecules and Ions to Be Investigated | | |
|---|---|---|
| **The Central Atom *Does* Obey the Octet Rule** | | |
| $NH_3$ (example) | HBr | $NO_3^-$ |
| $SiH_4$ | CHBrClF | $NH_4^+$ |
| CO | $H_2S$ | $CO_3^{2-}$ |
| **The Central Atom *Does Not* Obey the Octet Rule** | | |
| $ClF_3$ | $PCl_5$ | $AlCl_3$ |
| $I_3^-$ | $SF_6$ | |
| **This Multiple-Centered Molecule *Does* Obey the Octet Rule** (and duet rule for H) | | |
| $CH_3COOH$ | | |

# The Central Atom in Species 1–8 *Does* Obey the Octet Rule

**1. Silicon Tetrahydride: SiH$_4$**

| TABLE 3 | | | | | | |
| --- | --- | --- | --- | --- | --- | --- |
| List of Atoms | Number of Atoms | Valence Electrons per Atom | Total Valence Electrons for These Atoms | Ion? (add electrons for negative ion; subtract electrons for positive ion) | Total Valence Electrons (add all of the valence electrons and ion electrons) | Total Number of Bonds and Lone Pairs of Electrons |
| Si | | | | | | |
| H | | | | | | |

**a.** Draw the Lewis structure:

**b.** Report the number of bonding electron pairs on the central atom:

**c.** Report the number of non-bonding electron pairs on the central atom:

**d.** Report the total number of electron pairs on the central atom:

**e.** Name the electron set distribution about the central atom:

**f.** Name the geometrical shape for this molecule:

**g.** Make a model of the molecule, have your TA approve it, and then sketch the 3D model below using dashes and wedges where necessary.

TA approval:

## 2. Carbon Monoxide: CO

| TABLE 4 | | | | | | |
|---|---|---|---|---|---|---|
| List of Atoms | Number of Atoms | Valence Electrons per Atom | Total Valence Electrons for These Atoms | Ion? (add electrons for negative ion; subtract electrons for positive ion) | Total Valence Electrons (add all of the valence electrons and ion electrons) | Total Number of Bonds and Lone Pairs of Electrons |
| C | | | | | | |
| O | | | | | | |

**a.** Draw the Lewis structure:

**b.** Report the number of bonding electron pairs on: C =              , O =

**c.** Report the number of non-bonding electron pairs on: C =            , O =

**d.** Report the total number of electron pairs on: C =            , O =

**e.** Report the **number of electron sets** on: C =            , O =

**f.** Name the electron set distribution about: C =            , O =

**g.** Name the geometrical shape for this diatomic molecule:

**h.** Make a model of the molecule, have your TA approve it, and then sketch the 3D model below using dashes and wedges where necessary.

TA approval:

### 3. Hydrogen Bromide: HBr

| TABLE 5 | | | | | | |
|---|---|---|---|---|---|---|
| List of Atoms | Number of Atoms | Valence Electrons per Atom | Total Valence Electrons for These Atoms | Ion? (add electrons for negative ion; subtract electrons for positive ion) | Total Valence Electrons (add all of the valence electrons and ion electrons) | Total Number of Bonds and Lone Pairs of Electrons |
| H | | | | | | |
| Br | | | | | | |

**a.** Draw the Lewis structure:

**b.** Report the number of bonding electron pairs on the bromine atom:

**c.** Report the number of non-bonding electron pairs on the bromine atom:

**d.** Report the total number of electron pairs on the bromine atom:

**e.** Name the electron set distribution about the bromine atom:

**f.** Name the geometrical shape for this diatomic molecule:

**g.** Make a model of the molecule, have your TA approve it, and then sketch the 3D model below using dashes and wedges where necessary.

TA approval:

### 4. Bromochlorofluoromethane: CHBrClF

| TABLE 6 | | | | | | |
| --- | --- | --- | --- | --- | --- | --- |
| List of Atoms | Number of Atoms | Valence Electrons per Atom | Total Valence Electrons for These Atoms | Ion? (add electrons for negative ion; subtract electrons for positive ion) | Total Valence Electrons (add all of the valence electrons and ion electrons) | Total Number of Bonds and Lone Pairs of Electrons |
| C | | | | | | |
| H | | | | | | |
| Br | | | | | | |
| Cl | | | | | | |
| F | | | | | | |

**a.** Draw the Lewis structure:

**b.** Report the number of bonding electron pairs on the central atom:

**c.** Report the number of non-bonding electron pairs on the central atom:

**d.** Report the total number of electron pairs on the central atom:

**e.** Name the electron set distribution about the central atom:

**f.** Name the geometrical shape for this molecule:

**g.** Make a model of the molecule, have your TA approve it, and then sketch the 3D model below using dashes and wedges where necessary.

TA approval:

## 5. Dihydrogen Sulfide: $H_2S$

**TABLE 7**

| List of Atoms | Number of Atoms | Valence Electrons per Atom | Total Valence Electrons for These Atoms | Ion? (add electrons for negative ion; subtract electrons for positive ion) | Total Valence Electrons (add all of the valence electrons and ion electrons) | Total Number of Bonds and Lone Pairs of Electrons |
|---|---|---|---|---|---|---|
| H | | | | | | |
| S | | | | | | |

**a.** Draw the Lewis structure:

**b.** Report the number of bonding electron pairs on the central atom:

**c.** Report the number of non-bonding electron pairs on the central atom:

**d.** Report the total number of electron pairs on the central atom:

**e.** Name the electron set distribution about the central atom:

**f.** Report the **number of bonding electron sets** on the central atom:

**g.** Report the **number of non-bonding electron sets** on the central atom:

**h.** Name the geometrical shape for this molecule:

**i.** Make a model of the molecule, have your TA approve it, and then sketch the 3D model below using dashes and wedges where necessary.

TA approval:

## 6. Nitrate Anion: $NO_3^-$

**TABLE 8**

| List of Atoms | Number of Atoms | Valence Electrons per Atom | Total Valence Electrons for These Atoms | Ion? (add electrons for negative ion; subtract electrons for positive ion) | Total Valence Electrons (add all of the valence electrons and ion electrons) | Total Number of Bonds and Lone Pairs of Electrons |
|---|---|---|---|---|---|---|
| N | | | | | | |
| O | | | | | | |

**a.** Draw the Lewis structure:

**b.** Report the number of bonding electron pairs on the central atom:

**c.** Report the number of non-bonding electron pairs on the central atom:

**d.** Report the total number of electron pairs on the central atom:

**e.** Report the **number of bonding electron sets** on the central atom:

**f.** Report the **number of non-bonding electron sets** about the central atom:

**g.** Name the electron set distribution about the central atom:

**h.** Name the geometrical shape for this anion:

**i.** Make a model of the anion, have your TA approve it, and then sketch the 3D model below using dashes and wedges where necessary.

TA approval:

### 7. Ammonium Cation: $NH_4^+$

**TABLE 9**

| List of Atoms | Number of Atoms | Valence Electrons per Atom | Total Valence Electrons for These Atoms | Ion? (add electrons for negative ion; subtract electrons for positive ion) | Total Valence Electrons (add all of the valence electrons and ion electrons) | Total Number of Bonds and Lone Pairs of Electrons |
|---|---|---|---|---|---|---|
| N | | | | | | |
| H | | | | | | |

**a.** Draw the Lewis structure:

**b.** Report the number of bonding electron pairs on the central atom:

**c.** Report the number of non-bonding electron pairs on the central atom:

**d.** Report the total number of electron pairs on the central atom:

**e.** Name the electron set distribution about the central atom:

**f.** Name the geometrical shape for this cation:

**g.** Make a model of the cation, have your TA approve it, and then sketch the 3D model below using dashes and wedges where necessary.

TA approval:

## 8. Carbonate Anion: $CO_3^{2-}$

| | | | | | | |
|---|---|---|---|---|---|---|
| **TABLE 10** | | | | | | |
| **List of Atoms** | **Number of Atoms** | **Valence Electrons per Atom** | **Total Valence Electrons for These Atoms** | **Ion?** (add electrons for negative ion; subtract electrons for positive ion) | **Total Valence Electrons** (add all of the valence electrons and ion electrons) | **Total Number of Bonds and Lone Pairs of Electrons** |
| C | | | | | | |
| O | | | | | | |

**a.** Draw the Lewis structure:

**b.** Report the number of bonding electron pairs on the central atom:

**c.** Report the number of non-bonding electron pairs on the central atom:

**d.** Report the total number of electron pairs on the central atom:

**e.** Report the **number of electron sets** on the central atom:

**f.** Name the electron set distribution about the central atom:

**g.** Name the geometrical shape for this anion:

**h.** Make a model of the anion, have your TA approve it, and then sketch the 3D model below using dashes and wedges where necessary.

TA approval:

# The Central Atom in Species 9–13
## *Does Not* Obey the Octet Rule

**9. Chlorine TriFluoride: $ClF_3$**

| TABLE 11 | | | | | | |
|---|---|---|---|---|---|---|
| List of Atoms | Number of Atoms | Valence Electrons per Atom | Total Valence Electrons for These Atoms | Ion? (add electrons for negative ion; subtract electrons for positive ion) | Total Valence Electrons (add all of the valence electrons and ion electrons) | Total Number of Bonds and Lone Pairs of Electrons |
| Cl | | | | | | |
| F | | | | | | |

**a.** Draw the Lewis structure:

**b.** Report the number of bonding electron pairs on the central atom:

**c.** Report the number of non-bonding electron pairs on the central atom:

**d.** Report the total number of electron pairs on the central atom:

**e.** Name the electron set distribution about the central atom:

**f.** Name the geometrical shape for this molecule:

**g.** Make a model of the molecule, have your TA approve it, and then sketch the 3D model below using dashes and wedges where necessary.

TA approval:

## 10. Triiodide: $I_3^-$

| TABLE 12 | | | | | | |
| --- | --- | --- | --- | --- | --- | --- |
| List of Atoms | Number of Atoms | Valence Electrons per Atom | Total Valence Electrons for These Atoms | Ion? (add electrons for negative ion; subtract electrons for positive ion) | Total Valence Electrons (add all of the valence electrons and ion electrons) | Total Number of Bonds and Lone Pairs of Electrons |
| I | | | | | | |

**a.** Draw the Lewis structure:

**b.** Report the number of bonding electron pairs on the central atom:

**c.** Report the number of non-bonding electron pairs on the central atom:

**d.** Report the total number of electron pairs on the central atom =

**e.** Name the electron set distribution about the central atom:

**f.** Name the geometrical shape for this anion:

**g.** Make a model of the anion, have your TA approve it, and then sketch the 3D model below using dashes and wedges where necessary.

TA approval:

## 11. Phosphorous Pentachloride: $PCl_5$

**TABLE 13**

| List of Atoms | Number of Atoms | Valence Electrons per Atom | Total Valence Electrons for These Atoms | Ion? (add electrons for negative ion; subtract electrons for positive ion) | Total Valence Electrons (add all of the valence electrons and ion electrons) | Total Number of Bonds and Lone Pairs of Electrons |
|---|---|---|---|---|---|---|
| P | | | | | | |
| Cl | | | | | | |

**a.** Draw the Lewis structure:

**b.** Report the number of bonding electron pairs on the central atom:

**c.** Report the number of non-bonding electron pairs on the central atom:

**d.** Report the total number of electron pairs on the central atom:

**e.** Name the electron set distribution about the central atom:

**f.** Name the geometrical shape for this molecule:

**g.** Make a model of the molecule, have your TA approve it, and then sketch the 3D model below using dashes and wedges where necessary.

TA approval:

## 12. Sulfur Hexafluoride: $SF_6$

**TABLE 14**

| List of Atoms | Number of Atoms | Valence Electrons per Atom | Total Valence Electrons for These Atoms | Ion? (add electrons for negative ion; subtract electrons for positive ion) | Total Valence Electrons (add all of the valence electrons and ion electrons) | Total Number of Bonds and Lone Pairs of Electrons |
|---|---|---|---|---|---|---|
| S | | | | | | |
| F | | | | | | |

   **a.** Draw the Lewis structure:

   **b.** Report the number of bonding electron pairs on the central atom:

   **c.** Report the number of non-bonding electron pairs on the central atom:

   **d.** Report the total number of electron pairs on the central atom:

   **e.** Name the electron set distribution about the central atom:

   **f.** Name the geometrical shape for this molecule:

   **g.** Make a model of the molecule, have your TA approve it, and then sketch the 3D model below using dashes and wedges where necessary.

   TA approval:

### 13. Aluminum Chloride: $AlCl_3$

| TABLE 15 | | | | | | |
|---|---|---|---|---|---|---|
| List of Atoms | Number of Atoms | Valence Electrons per Atom | Total Valence Electrons for These Atoms | Ion? (add electrons for negative ion; subtract electrons for positive ion) | Total Valence Electrons (add all of the valence electrons and ion electrons) | Total Number of Bonds and Lone Pairs of Electrons |
| Al | | | | | | |
| Cl | | | | | | |

  **a.** Draw the Lewis structure:

  **b.** Report the number of bonding electron pairs on the central atom:

  **c.** Report the number of non-bonding electron pairs on the central atom:

  **d.** Report the total number of electron pairs on the central atom:

  **e.** Name the electron set distribution about the central atom:

  **f.** Name the geometrical shape for this molecule:

  **g.** Make a model of the molecule, have your TA approve it, and then sketch the 3D model below using dashes and wedges where necessary.

  TA approval:

# Species 14 *Does* Obey the Octet Rule (and duet rule for H)

**14. Acetic Acid: $CH_3COOH$**

**TABLE 16**

| List of Atoms | Number of Atoms | Valence Electrons per Atom | Total Valence Electrons for These Atoms | Ion? (add electrons for negative ion; subtract electrons for positive ion) | Total Valence Electrons (add all of the valence electrons and ion electrons) | Total Number of Bonds and Lone Pairs of Electrons |
|---|---|---|---|---|---|---|
| C | | | | | | |
| H | | | | | | |
| O | | | | | | |

**a.** Draw the Lewis structure:

**b.** Report the number of bonding electron pairs on the carbon atom bonded to three Hs:

**c.** Report the number of bonding electron pairs on the carbon bonded to two Os:

**d.** Report the number of bonding electron pairs on the O bonded H ("hydroxyl O"):

**e.** Report the number of non-bonding electron pairs on the "hydroxyl O":

**f.** Report the number of non-bonding electron pairs on each carbon atom: (the same for each)

**g.** Report the total number of electron pairs on each carbon atom = (the same for each again)

**h.** Name the electron set distribution about the carbon atom bonded to three Hs (the "methyl" carbon) *and* the geometry about this "methyl" carbon atom:

*and*

**i.** Re-draw your valid Lewis structure below for reference:

**j.** Report the **number of bonding electron sets** on the carbon atom bonded to two Os:

**k.** Report the **number of non-bonding electron sets** on the carbon bonded to two Os:

**l.** Name the electron set distribution about the carbon atom bonded to two Os (the "carboxyl" carbon) **and** the geometry about this "carboxyl" carbon atom:

**and**

**m.** Name the electron set distribution about the "hydroxyl oxygen" atom **and** the geometry about the "hydroxyl oxygen" atom:

**and**

**n.** Make a model of the molecule, have your TA approve it, and then sketch the 3D model below using dashes and wedges where necessary.

TA approval:

# Formal Charge and Atomic Orbital Hybridization of Molecules and Polyatomic Ions



## CHE 112L

- An adaptation and extension of the CHE 111L Lewis Electron Dot Diagram (LEDD) laboratory exercise

# BACKGROUND

**TABLE 1  The Shapes of Molecules**

| Atomic Orbital Hybridization Name (number of groups) | sp-hybridized (2) | sp²-hybridized (3) | | sp³-hybridized (4) | | |
|---|---|---|---|---|---|---|
| **Molecular Shape (class)** | Linear ($AX_2$) | Trigonal Planar ($AX_3$) | V-Shaped or Bent ($AX_2E$) | Tetrahedral ($AX_4$) | Trigonal Pyramidal ($AX_3E$) | V-Shaped or Bent ($AX_2E_2$) |
| **Number of Bonding Groups** | 2 | 3 | 2 | 4 | 3 | 2 |
| **Number of Lone Pairs** | 0 | 0 | 1 | 0 | 1 | 2 |
| **Bond Angle** | 180° | 120° | <120° | 109.5° | <109.5° | <109.5° |

| Atomic Orbital Hybridization Name (number of groups) | sp³d-hybridized (5) | | | | sp³d²-hybridized (6) | | |
|---|---|---|---|---|---|---|---|
| **Molecular Shape (class)** | Trigonal Bypyramidal ($AX_5$) | Seesaw ($AX_4E$) | T-Shaped ($AX_3E_2$) | Linear ($AX_2E_3$) | Octahedral ($AX_6$) | Square Pyramidal ($AX_5E$) | Square Planar ($AX_4E_2$) |
| **Number of Bonding Groups** | 5 | 4 | 3 | 2 | 6 | 5 | 4 |
| **Number of Lone Pairs** | 0 | 1 | 2 | 3 | 0 | 1 | 2 |
| **Bond Angle** | 90° (ax) 120° (eq) | <90° (ax) <120° (eq) | <90° (ax) | 180° | 90° | <90° | 90° |

# DATA SHEET

Name: _________________________________    Grade: _________________________________

Date Experiment Performed: _______________    Days Late: _________________________

CRN of Lab Section: _____________________    Instructor's Initials: _____________________

| TABLE 2   The 14 Molecules and Ions to Be Investigated | | |
|---|---|---|
| **These Species *Do* Obey the Octet Rule** | | |
| $SiH_4$ | CHBrClF | $NH_4^+$ |
| CO | $H_2S$ | $CO_3^{2-}$ |
| HBr | $NO_3^-$ | |
| **These Species *Do Not* Obey the Octet Rule** | | |
| $ClF_3$ | $PCl_5$ | $AlCl_3$ |
| $I_3^-$ | $SF_6$ | |
| **This Multiple-Centered Molecule *Does* Obey the Octet Rule** | | |
| $CH_3COOH$ | | |

# Species 1–8 *Do* Obey the Octet Rule

**1. Silicon Tetrahydride: SiH$_4$**

   **a.** Draw the Lewis structure:

   **b.** Calculate the formal charge for each different type of atom in the Lewis structure in the space below. Be sure that the sum of the formal charges is equal to the overall charge on the molecule.

   **c.** What is the electron group arrangement about the central atom according to VSEPR theory and Figure 10.2, page 418 in Silberberg and Amateis? Draw a 2D representation of the molecule in the space below using wedges and dashes appropriately.

   **d.** Name the molecular shape of this molecule: 

      (Figure 10.10, page 424 in Silberberg and Amateis)

   **e.** What is the hybridization of the atomic orbitals on the central atom? 

      (Table 11.1, page 449 in Silberberg and Amateis)

## 2. Carbon Monoxide: CO

**a.** Draw the Lewis structure:

**b.** Calculate the formal charge for each different type of atom in the Lewis structure in the space below. Be sure that the sum of the formal charges is equal to the overall charge on the molecule.

**c.** What is the electron group arrangement about the central atom according to VSEPR theory and Figure 10.2, page 418 in Silberberg and Amateis? Draw a 2D representation of the molecule in the space below using wedges and dashes appropriately.

**d.** Name the molecular shape of this molecule:

(Figure 10.10, page 424 in Silberberg and Amateis)

**e.** What is the hybridization of the atomic orbitals on the central atom?

(Table 11.1, page 449 in Silberberg and Amateis)

### 3. Hydrogen Bromide: HBr

**a.** Draw the Lewis structure:

**b.** Calculate the formal charge for each different type of atom in the Lewis structure in the space below. Be sure that the sum of the formal charges is equal to the overall charge on the molecule.

**c.** What is the electron group arrangement about the central atom according to VSEPR theory and Figure 10.2, page 418 in Silberberg and Amateis? Draw a 2D representation of the molecule in the space below using wedges and dashes appropriately.

**d.** Name the molecular shape of this molecule:

(Figure 10.10, page 424 in Silberberg and Amateis)

**e.** What is the hybridization of the atomic orbitals on the central atom?

(Table 11.1, page 449 in Silberberg and Amateis)

**4. Bromochlorofluoromethane: CHBrClF**

   **a.** Draw the Lewis structure:

   **b.** Calculate the formal charge for each different type of atom in the Lewis structure in the space below. Be sure that the sum of the formal charges is equal to the overall charge on the molecule.

   **c.** What is the electron group arrangement about the central atom according to VSEPR theory and Figure 10.2, page 418 in Silberberg and Amateis? Draw a 2D representation of the molecule in the space below using wedges and dashes appropriately.

   **d.** Name the molecular shape of this molecule:

   (Figure 10.10, page 424 in Silberberg and Amateis)

   **e.** What is the hybridization of the atomic orbitals on the central atom?

   (Table 11.1, page 449 in Silberberg and Amateis)

**5. Dihydrogen Sulfide: $H_2S$**

    **a.** Draw the Lewis structure:

    **b.** Calculate the formal charge for each different type of atom in the Lewis structure in the space below. Be sure that the sum of the formal charges is equal to the overall charge on the molecule.

    **c.** What is the electron group arrangement about the central atom according to VSEPR theory and Figure 10.2, page 418 in Silberberg and Amateis? Draw a 2D representation of the molecule in the space below using wedges and dashes appropriately.

    **d.** Name the molecular shape of this molecule:

    (Figure 10.10, page 424 in Silberberg and Amateis)

    **e.** What is the hybridization of the atomic orbitals on the central atom?

    (Table 11.1, page 449 in Silberberg and Amateis)

**6. Nitrate Anion: $NO_3^-$**

   **a.** Draw the Lewis structure:

   **b.** Calculate the formal charge for each different type of atom in the Lewis structure in the space below. Be sure that the sum of the formal charges is equal to the overall charge on the molecule.

   **c.** What is the electron group arrangement about the central atom according to VSEPR theory and Figure 10.2, page 418 in Silberberg and Amateis? Draw a 2D representation of the molecule in the space below using wedges and dashes appropriately.

   **d.** Name the molecular shape of this molecule:

     (Figure 10.10, page 424 in Silberberg and Amateis)

   **e.** What is the hybridization of the atomic orbitals on the central atom?

     (Table 11.1, page 449 in Silberberg and Amateis)

**7. Ammonium Cation: $NH_4^+$**

    **a.** Draw the Lewis structure:

    **b.** Calculate the formal charge for each different type of atom in the Lewis structure in the space below. Be sure that the sum of the formal charges is equal to the overall charge on the molecule.

    **c.** What is the electron group arrangement about the central atom according to VSEPR theory and Figure 10.2, page 418 in Silberberg and Amateis? Draw a 2D representation of the molecule in the space below using wedges and dashes appropriately.

    **d.** Name the molecular shape of this molecule:

    (Figure 10.10, page 424 in Silberberg and Amateis)

    **e.** What is the hybridization of the atomic orbitals on the central atom?

    (Table 11.1, page 449 in Silberberg and Amateis)

**8. Carbonate Anion: $CO_3{}^{2-}$**

   **a.** Draw the Lewis structure:

   **b.** Calculate the formal charge for each different type of atom in the Lewis structure in the space below. Be sure that the sum of the formal charges is equal to the overall charge on the molecule.

   **c.** What is the electron group arrangement about the central atom according to VSEPR theory and Figure 10.2, page 418 in Silberberg and Amateis? Draw a 2D representation of the molecule in the space below using wedges and dashes appropriately.

   **d.** Name the molecular shape of this molecule:

     (Figure 10.10, page 424 in Silberberg and Amateis)

   **e.** What is the hybridization of the atomic orbitals on the central atom?

     (Table 11.1, page 449 in Silberberg and Amateis)

# Species 9–13 *Do Not* Obey the Octet Rule

**9. Chlorine TriFluoride: $ClF_3$**

**a.** Draw the Lewis structure:

**b.** Calculate the formal charge for each different type of atom in the Lewis structure in the space below. Be sure that the sum of the formal charges is equal to the overall charge on the molecule.

**c.** What is the electron group arrangement about the central atom according to VSEPR theory and Figure 10.2, page 418 in Silberberg and Amateis? Draw a 2D representation of the molecule in the space below using wedges and dashes appropriately.

**d.** Name the molecular shape of this molecule:

(Figure 10.10, page 424 in Silberberg and Amateis)

**e.** What is the hybridization of the atomic orbitals on the central atom?

(Table 11.1, page 449 in Silberberg and Amateis)

**10. Triiodide Anion: $I_3^-$**

    **a.** Draw the Lewis structure:

    **b.** Calculate the formal charge for each different type of atom in the Lewis structure in the space below. Be sure that the sum of the formal charges is equal to the overall charge on the molecule.

    **c.** What is the electron group arrangement about the central atom according to VSEPR theory and Figure 10.2, page 418 in Silberberg and Amateis? Draw a 2D representation of the molecule in the space below using wedges and dashes appropriately.

    **d.** Name the molecular shape of this molecule:

    (Figure 10.10, page 424 in Silberberg and Amateis)

    **e.** What is the hybridization of the atomic orbitals on the central atom?

    (Table 11.1, page 449 in Silberberg and Amateis)

**11. Phosphorous Pentachloride: PCl$_5$**

   **a.** Draw the Lewis structure:

   **b.** Calculate the formal charge for each different type of atom in the Lewis structure in the space below. Be sure that the sum of the formal charges is equal to the overall charge on the molecule.

   **c.** What is the electron group arrangement about the central atom according to VSEPR theory and Figure 10.2, page 418 in Silberberg and Amateis? Draw a 2D representation of the molecule in the space below using wedges and dashes appropriately.

   **d.** Name the molecular shape of this molecule:

   (Figure 10.10, page 424 in Silberberg and Amateis)

   **e.** What is the hybridization of the atomic orbitals on the central atom?

   (Table 11.1, page 449 in Silberberg and Amateis)

**12. Sulfur Hexafluoride: SF$_6$**

   **a.** Draw the Lewis structure:

   **b.** Calculate the formal charge for each different type of atom in the Lewis structure in the space below. Be sure that the sum of the formal charges is equal to the overall charge on the molecule.

   **c.** What is the electron group arrangement about the central atom according to VSEPR theory and Figure 10.2, page 418 in Silberberg and Amateis? Draw a 2D representation of the molecule in the space below using wedges and dashes appropriately.

   **d.** Name the molecular shape of this molecule:

   (Figure 10.10, page 424 in Silberberg and Amateis)

   **e.** What is the hybridization of the atomic orbitals on the central atom?

   (Table 11.1, page 449 in Silberberg and Amateis)

**13. Aluminum Chloride: $AlCl_3$**

    **a.** Draw the Lewis structure:

    **b.** Calculate the formal charge for each different type of atom in the Lewis structure in the space below. Be sure that the sum of the formal charges is equal to the overall charge on the molecule.

    **c.** What is the electron group arrangement about the central atom according to VSEPR theory and Figure 10.2, page 418 in Silberberg and Amateis? Draw a 2D representation of the molecule in the space below using wedges and dashes appropriately.

    **d.** Name the molecular shape of this molecule:

        (Figure 10.10, page 424 in Silberberg and Amateis)

    **e.** What is the hybridization of the atomic orbitals on the central atom?

        (Table 11.1, page 449 in Silberberg and Amateis)

# Species 14 *Does* Obey the Octet Rule

**14. Acetic Acid: CH$_3$COOH (does obey the octet and duet rules)**

   **a.** Draw the Lewis structure:

   **b.** Calculate the formal charge for each different type of atom in the Lewis structure in the space below. Be sure that the sum of the formal charges is equal to the overall charge on the molecule.

   **c.** According to VSEPR theory and Figure 10.2, page 418 in Silberberg and Amateis, what is the electron group arrangement about *each carbon atom* and about the *oxygen atom that the acidic hydrogen is bonded to?* Draw a 2D representation of the molecule in the space below using wedges and dashes appropriately.

**d.** State the "molecular geometry" about each carbon atom and about the oxygen atom that the acidic hydrogen is bonded to. (Figure 10.10, page 424 in Silberberg and Amateis)

**e.** What is the hybridization of the atomic orbitals on each carbon atom and on the oxygen atom that the acidic hydrogen is bonded to? (Table 11.1, page 449 in Silberberg and Amateis)

# Polarity and Intermolecular Forces of Molecules and Polyatomic Ions

## CHE 112L

- An adaptation and extension of the CHE 111L Molecular Geometry laboratory exercise

# 20-QUESTION PRE-LAB ASSIGNMENT

CRN:_______________________________________     Your Name: _______________________________________

Date Submitted: _______________________________     TA's Name: _______________________________________

*Please write your answers legibly in the space provided. Some answers will actually consist of three or four pieces of information that are clearly written on the board during the conversation **or** clearly stated verbally during the presentation. You must include all aspects of the answer to receive full credit. These answers are due at the beginning of the lab meeting.*

1.  What is the "Big Bird-style word" that you can use in lieu of Figure 9.5 on page 296 in the textbook? Include the numerical electronegativity values with each chemical symbol.

2.  What is the equation we use to calculate the electronegativity difference? Will the calculated value be always positive or always negative?

3.  Draw out the numerical "scheme" showing the ranges of values for polar versus non-polar and ionic versus covalent bonds.

4.  Because all ionic bonds are ________________, our emphasis in this lab activity is on ________________ or ________________ bonds to get you ready and thinking "112-like."

5.  If we have a polar (covalent) bond, what do we use to represent that?

**6.** Which two pieces of information combine to allow us to predict molecular polarity?

**7.** What can we talk about next after we have predicted the molecular polarity (or non-polarity)?

**8.** What are the three classic van der Waals forces (from the old days)?

**9.** What is the fourth force that your textbook authors include?

**10.** What are the five example molecules that we use? (List them left to right below.)

______ , ______ , ______ , ______ , ______ .

**11.** What are the types of chemical bonds in each? (List them left to right below.)

______ , ______ , ______ , ______ , ______ .

**12.** What is the electronegativity for each atom? (List them left to right below.)

______ , ______ , ______ , ______ , ______ .

**13.** What is the electronegativity difference for each molecule? (List them left to right below.)

______ , ______ , ______ , ______ , ______ .

**14.** If a bond dipole exists, show it. Otherwise write "none." (Show them left to right below.)

**15.** Use the dipole moments to represent the molecule. (Show them left to right below.)

**16.** Finally, tell whether the molecule is polar or not. (List them left to right below.)

_______________ , _______________ , _______________ , _______________ , _______________ .

**17.** What did we call the ion-dipole interaction in CHE 111L? Draw out the CHE 111L representation **and** the CHE 112L representation shown in the video.

**18.** What is an abbreviated designation for the requirements of the hydrogen bond? List four examples of molecules that can hydrogen bond.

**19.** What is necessary for dipole–dipole interactions? **And** give examples from the video.

**20.** What does the London dispersive force require?

# DATA SHEET

Name: _________________________________   Grade: _________________________________

Date Experiment Performed: _______________   Days Late: _______________________________

CRN of Lab Section: _____________________   Instructor's Initials: _____________________

| TABLE 1   The 14 Molecules and Ions to Be Investigated | | |
|---|---|---|
| **These Species *Do* Obey the Octet Rule** | | |
| $SiH_4$ | CHBrClF | $NH_4^+$ |
| CO | $H_2S$ | $CO_3^{2-}$ |
| HBr | $NO_3^-$ | |
| **These Species *Do Not* Obey the Octet Rule** | | |
| $ClF_3$ | $PCl_5$ | $AlCl_3$ |
| $I_3^-$ | $SF_6$ | |
| **This Multiple-Centered Molecule *Does* Obey the Octet Rule** | | |
| $CH_3COOH$ | | |

# Species 1–8 *Do* Obey the Octet Rule

**1. Silicon Tetrahydride: SiH$_4$**

    **a.** List the bonds that exist between the central atom and each peripheral atom. (For example, O–H in the water molecule, N–H in the ammonia molecule, etc.)

    **b.** For each atom in this molecule, obtain the numerical electronegativity value from Figure 9.21, page 391 in the textbook, and write it below.

    **c.** Calculate the electronegativity difference for each bond. State whether each bond is polar or non-polar.

    **d.** Draw the geometry of this molecule using dipole moments.

    **e.** You predict this molecule to be (circle one):       Polar       Non-polar

    **f.** List *all* the intermolecular forces that can exist for this molecule and select the dominant IMF.

**2. Carbon Monoxide: CO**

   **a.** List the bonds that exist between the central atom and each peripheral atom. (For example, O–H in the water molecule, N–H in the ammonia molecule, etc.)

   **b.** For each atom in this molecule, obtain the numerical electronegativity value from Figure 9.21, page 391 in the textbook, and write it below.

   **c.** Calculate the electronegativity difference for each bond. State whether each bond is polar or non-polar.

   **d.** Draw the geometry of this molecule using dipole moments.

   **e.** You predict this molecule to be (circle one):      Polar        Non-polar

   **f.** List *all* the intermolecular forces that can exist for this molecule and select the dominant IMF.

**3. Hydrogen Bromide: HBr**

  **a.** List the bonds that exist between the central atom and each peripheral atom. (For example, O–H in the water molecule, N–H in the ammonia molecule, etc.)

  **b.** For each atom in this molecule, obtain the numerical electronegativity value from Figure 9.21, page 391 in the textbook, and write it below.

  **c.** Calculate the electronegativity difference for each bond. State whether each bond is polar or non-polar.

  **d.** Draw the geometry of this molecule using dipole moments.

  **e.** You predict this molecule to be (circle one):          Polar          Non-polar

  **f.** List **all** the intermolecular forces that can exist for this molecule and select the dominant IMF.

**4. Bromochlorofluoromethane: CHBrClF**

**a.** List the bonds that exist between the central atom and each peripheral atom. (For example, O–H in the water molecule, N–H in the ammonia molecule, etc.)

**b.** For each atom in this molecule, obtain the numerical electronegativity value from Figure 9.21, page 391 in the textbook, and write it below.

**c.** Calculate the electronegativity difference for each bond. State whether each bond is polar or non-polar.

**d.** Draw the geometry of this molecule using dipole moments.

**e.** You predict this molecule to be (circle one):          Polar          Non-polar

**f.** List *all* the intermolecular forces that can exist for this molecule and select the dominant IMF.

**5. Dihydrogen Sulfide: $H_2S$**

    **a.** List the bonds that exist between the central atom and each peripheral atom. (For example, O–H in the water molecule, N–H in the ammonia molecule, etc.)

    **b.** For each atom in this molecule, obtain the numerical electronegativity value from Figure 9.21, page 391 in the textbook, and write it below.

    **c.** Calculate the electronegativity difference for each bond. State whether each bond is polar or non-polar.

    **d.** Draw the geometry of this molecule using dipole moments.

    **e.** You predict this molecule to be (circle one):        Polar        Non-polar

    **f.** List **all** the intermolecular forces that can exist for this molecule and select the dominant IMF.

**6. Nitrate Anion: $NO_3^-$**

   **a.** List the bonds that exist between the central atom and each peripheral atom. (For example, O–H in the water molecule, N–H in the ammonia molecule, etc.)

   **b.** For each atom in this polyatomic anion, obtain the numerical electronegativity value from Figure 9.21, page 391 in the textbook, and write it below.

   **c.** Calculate the electronegativity difference for each bond. State whether each bond is polar or non-polar.

   **d.** Draw the geometry of this polyatomic anion using dipole moments.

   **e.** You predict this polyatomic anion to be (circle one):     Polar     Non-polar

   **f.** List **all** the intermolecular forces that can exist for this anion and select the dominant IMF.

**7. Ammonium Cation: $NH_4^+$**

    **a.** List the bonds that exist between the central atom and each peripheral atom. (For example, O–H in the water molecule, N–H in the ammonia molecule, etc.)

    **b.** For each atom in this polyatomic cation, obtain the numerical electronegativity value from Figure 9.21, page 391 in the textbook, and write it below.

    **c.** Calculate the electronegativity difference for each bond. State whether each bond is polar or non-polar.

    **d.** Draw the geometry of this polyatomic cation using dipole moments.

    **e.** You predict this polyatomic cation to be (circle one):    Polar    Non-polar

    **f.** List *all* the intermolecular forces that can exist for this cation and select the dominant IMF.

**8. Carbonate Anion: $CO_3{}^{2-}$**

    **a.** List the bonds that exist between the central atom and each peripheral atom. (For example, O–H in the water molecule, N–H in the ammonia molecule, etc.)

    **b.** For each atom in this polyatomic anion, obtain the numerical electronegativity value from Figure 9.21, page 391 in the textbook, and write it below.

    **c.** Calculate the electronegativity difference for each bond. State whether each bond is polar or non-polar.

    **d.** Draw the geometry of this polyatomic anion using dipole moments.

    **e.** You predict this polyatomic anion to be (circle one):    Polar    Non-polar

    **f.** List *all* the intermolecular forces that can exist for this anion and select the dominant IMF.

# Species 9–13 *Do Not* Obey the Octet Rule

**9. Chlorine TriFluoride: $ClF_3$**

   **a.** List the bonds that exist between the central atom and each peripheral atom. (For example, O–H in the water molecule, N–H in the ammonia molecule, etc.)

   **b.** For each atom in this molecule, obtain the numerical electronegativity value from Figure 9.21, page 391 in the textbook, and write it below.

   **c.** Calculate the electronegativity difference for each bond. State whether each bond is polar or non-polar.

   **d.** Draw the geometry of this molecule using dipole moments.

   **e.** You predict this molecule to be (circle one):        Polar        Non-polar

   **f.** List *all* the intermolecular forces that can exist for this molecule and select the dominant IMF.

**10. Triiodide Anion: $I_3^-$**

    **a.** List the bonds that exist between the central atom and each peripheral atom. (For example, O–H in the water molecule, N–H in the ammonia molecule, etc.)

    **b.** For each atom in this polyatomic anion, obtain the numerical electronegativity value from Figure 9.21, page 391 in the textbook, and write it below.

    **c.** Calculate the electronegativity difference for each bond. State whether each bond is polar or non-polar.

    **d.** Draw the geometry of this polyatomic anion using dipole moments.

    **e.** You predict this polyatomic anion to be (circle one):    Polar    Non-polar

    **f.** List ***all*** the intermolecular forces that can exist for this anion and select the dominant IMF.

**11. Phosphorous Pentachloride: $PCl_5$**

    **a.** List the bonds that exist between the central atom and each peripheral atom. (For example, O–H in the water molecule, N–H in the ammonia molecule, etc.)

    **b.** For each atom in this molecule, obtain the numerical electronegativity value from Figure 9.21, page 391 in the textbook, and write it below.

    **c.** Calculate the electronegativity difference for each bond. State whether each bond is polar or non-polar.

    **d.** Draw the geometry of this molecule using dipole moments.

    **e.** You predict this molecule to be (circle one):       Polar       Non-polar

    **f.** List *all* the intermolecular forces that can exist for this molecule and select the dominant IMF.

**12. Sulfur Hexafluoride: SF$_6$**

**a.** List the bonds that exist between the central atom and each peripheral atom. (For example, O–H in the water molecule, N–H in the ammonia molecule, etc.)

**b.** For each atom in this molecule, obtain the numerical electronegativity value from Figure 9.21, page 391 in the textbook, and write it below.

**c.** Calculate the electronegativity difference for each bond. State whether each bond is polar or non-polar.

**d.** Draw the geometry of this molecule using dipole moments.

**e.** You predict this molecule to be (circle one):       Polar       Non-polar

**f.** List **all** the intermolecular forces that can exist for this molecule and select the dominant IMF.

**13. Aluminum Chloride: $AlCl_3$**

    **a.** List the bonds that exist between the central atom and each peripheral atom. (For example, O–H in the water molecule, N–H in the ammonia molecule, etc.)

    **b.** For each atom in this molecule, obtain the numerical electronegativity value from Figure 9.21, page 391 in the textbook, and write it below.

    **c.** Calculate the electronegativity difference for each bond. State whether each bond is polar or non-polar.

    **d.** Draw the geometry of this molecule using dipole moments.

    **e.** You predict this molecule to be (circle one):       Polar       Non-polar

    **f.** List *all* the intermolecular forces that can exist for this molecule and select the dominant IMF.

# Species 14 *Does* Obey the Octet Rule

**14. Acetic Acid: CH$_3$COOH**

    **a.** List the bonds that exist between the central atom and each peripheral atom. (For example, O–H in the water molecule, N–H in the ammonia molecule, etc.)

    **b.** For each atom in this molecule, obtain the numerical electronegativity value from Figure 9.21, page 391 in the textbook, and write it below.

    **c.** Calculate the electronegativity difference for each bond. State whether each bond is polar or non-polar.

    **d.** Draw the geometry of this molecule using dipole moments.

    **e.** You predict this molecule to be (circle one):       Polar        Non-polar

    **f.** List *all* the intermolecular forces that can exist for this molecule and select the dominant IMF.

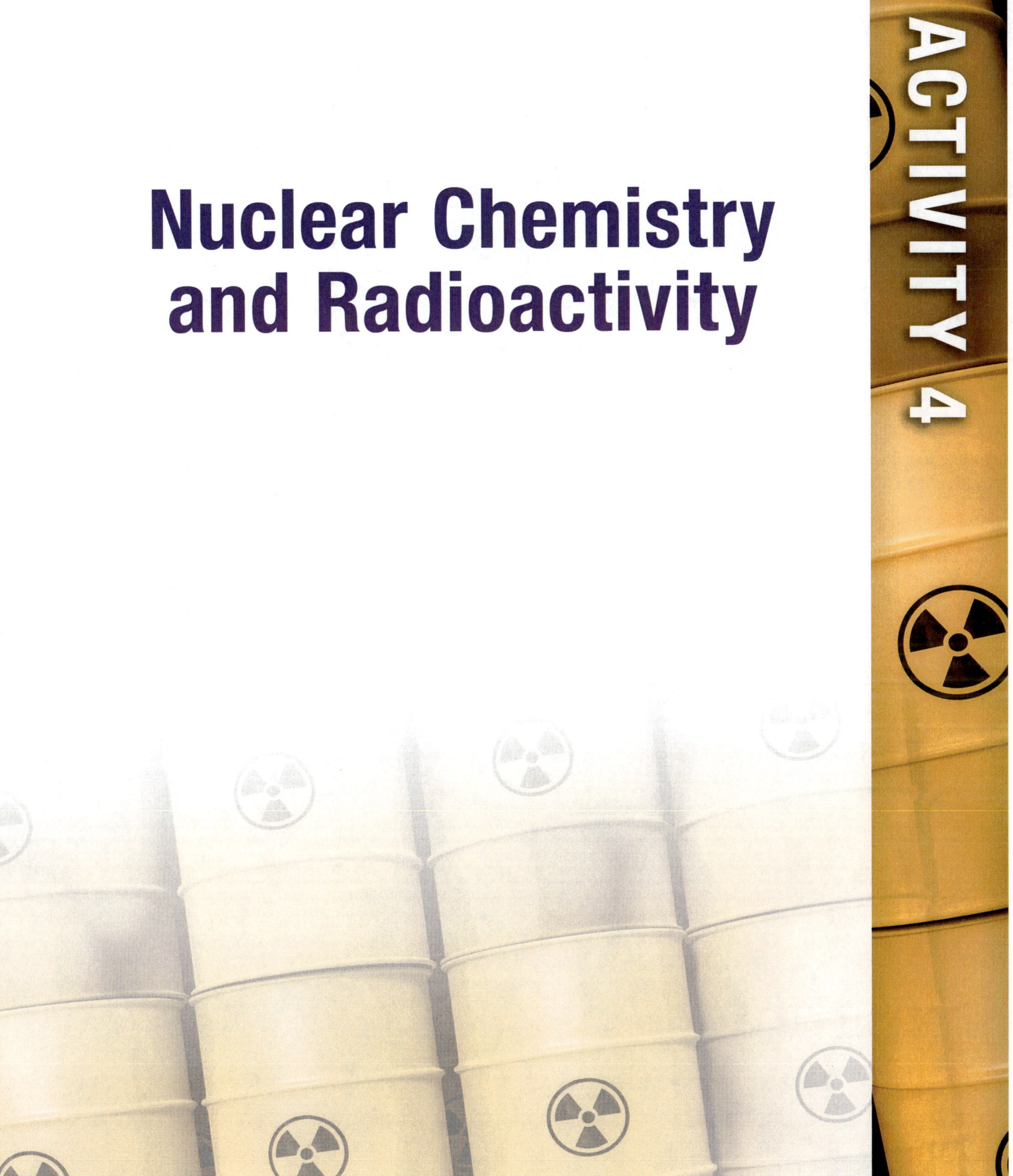

# Nuclear Chemistry and Radioactivity

# 20-QUESTION PRE-LAB ASSIGNMENT

CRN:_______________________________     Your Name: _______________________________

Date Submitted: _______________________     TA's Name: _______________________________

*Please write your answers legibly in the space provided.  Some answers will actually consist of three or four pieces of information that are clearly written on the board during the conversation **or** clearly stated verbally during the presentation. You must include all aspects of the answer to receive full credit. These answers are due at the beginning of the lab meeting.*

**1.** What is the text book definition for **radioactivity?**

**2.** What does **spontaneous** mean?

**3.** What *is* **radiation?**

**4.** What are the five types of radiation given off by atoms?

**5.** Write the more rigorous definition of **radioactivity.**

**6.** What is meant by **unstable** here?

**7.** Show an example of each—α, β, and γ radiation—for a **parent → daughter** decay process.

**8.** Which **law** of chemistry must be satisfied when a parent is converted to a daughter?

**9.** What is the definition of **chemistry** after Chapter 7?

**10.** Show how flerovium and livermorium were made by transmutation.

**11.** How were *all* of the transuranium elements made?

**12.** What other aspect of stability did we point out relating to Z, the atomic number?  What is the last stable isotope?

**13.** If *not* diamonds and rust, what *will be* left on earth at the end of time?

**14.** Define **half-life,** and include the symbol we use to represent it.

**15.** Show how 10 grams of **carbon-14*** decays down to 1.25 grams.

**16.** Draw a representation of the plot of **counts/minute versus time,** showing the shape of the plotted data and the proportionality upon which we depend.

**17.** Show how we can transform from a parabolic plot to a linear plot.

**18.** What is the useful relationship in the linear plot?

**19.** What is **non-spontaneous radioactivity** called, and what gets it started?

**20.** Show the **chain reaction** for uranium-235? How can we stop it (or at least slow it down)?

# BACKGROUND

Each element has a symbol such as C for carbon. However, sometimes simply a C is not sufficient to represent the atom of carbon being considered. Every atom of carbon has six protons
in its nucleus; atoms of carbon can have different numbers of neutrons and still be carbon.
Thus the symbols

$^{12}_{6}C$ and $^{13}_{6}C$

represent two different atoms of carbon, one with six neutrons and one with seven neutrons
respectively. The subscript is the number of protons in the nucleus (atomic number), and
the superscript is the sum of the number of *protons plus neutrons* (the mass number) in the
nucleus. These two atoms are said to be isotopes of carbon. An isotope has the same number
of protons (atomic number) but different number of neutrons (different mass number). The
carbon isotopes above can also be symbolized as C-12 and C-13. Not all elements have isotopes
that exist in nature, but counting the isotopes made by man, isotopes of all elements exist.
There are slightly over 100 elements, and there are about 3000 different atoms (nuclides) if
all the isotopes of all elements are added up.

Some nuclei of atoms are stable; that is, they remain unchanged with the passage of time.
Other nuclei are unstable and undergo transformation with the passage of time. These unstable nuclei are said to be radioactive, and they transform (emit particles from their nuclei).
Radioactive nuclei are sometimes called radioisotopes. When nuclei transform (decay),
they are converted from atoms of one element into atoms of another element. For instance,
the equation

$$^{238}_{92}U \rightarrow {}^{234}_{90}Th + {}^{4}_{2}He$$

indicates that a U-238 nucleus will transform to give a Th-234 nucleus and a helium nucleus
(a helium nucleus is called an alpha particle in this case). Thus, a uranium nucleus has been
converted to a thorium nucleus by radioactive transformation.

There are many nuclei that will transform as the uranium does; many others transform by
processes that involve emission of particles other than an alpha particle. The alpha particle,
other particles, and gamma rays (high energy light) emitted from radioisotopes are collec-
tively called radiation and can often be dangerous to human life or health. This radiation can
also be used to help human health or life.

No matter what the type of transformation, one important characteristic is the half-life of
the transformation. The half-life is the time required for one-half of the sample to transform.
If one starts with 4 grams of a uranium isotope, the time required for the sample to transform
to 2 grams of uranium is called the half-life for the uranium isotope. This amount of time is
the same for the 2 grams of the uranium to transform to 1 gram of uranium. No matter how
much sample one starts with, it requires the same time for half of it to transform.

Each radioactive isotope of any element has a half-life characteristic of that isotope.

Most radioisotopes are produced in nuclear reactors or cyclotrons. The study of the isotopes, their half-lives, and processes is a part of the field of nuclear chemistry. Besides its familiar use in nuclear reactors, the field of nuclear chemistry is encountered in such diverse areas as oil well logging, food preservation, basic research, and medicine.

Radioisotopes are used in diagnostic and therapeutic medicine for over 10% of the patients admitted to hospitals. Therapeutically, the radiation emitted by radioisotopes will in most cases preferentially destroy cancer cells in the body. The cancer cells are usually multiplying very rapidly and are not highly differentiated as healthy cells, which makes them more susceptible to radiation damage. However, most internally administered radioisotopes are used for diagnostic purposes rather than for therapeutic purposes.

Before these radioisotopes can be administered to humans, two very important questions have to be answered: (1) availability of radioisotopes, (2) effects of radiation exposure to the patient and others. Any radioisotope to be used in patients for diagnostic purposes should have a short half-life. The short half-life is necessary in order to reduce radiation exposure to the patient and also to produce better imagery in the instruments used in the diagnostic studies.

For example, iodide concentrates in the thyroid gland. When I-131 (which has a half-life of eight days) is administered to a patient thought to have a thyroid disorder, only a few hours are needed in order to determine the presences and location of a malignant tumor in the thyroids. However, sixty to seventy days are required for the patient's body to be free of the radioactive iodide either by secretions mainly in the urine or from the natural transformation of the radioactivity.

The ideal half-life of the radioactive iodide would be about four hours. If it takes more than thirty two hours to receive the radioisotope of iodide (I-31) from the nuclear reactor or cyclotron, all of the radioisotope will have decayed before it is received. The problem of quick availability must be considered for short half-life radioisotopes.

Until the radioactive cow was discovered, the use of short half-life radioisotopes in most hospitals was impractical because there are only a few nuclear reactors and cyclotron facilities. As you know, cows produce milk that can be removed by milking. A radioactive cow is a device that contains a radioisotope that transforms to produce a second radioisotope with a short half-life. The second radioisotope can be removed by "milking" the cow. In this situation, milking involves a separation of two elements that differ in some physical property such as solubility of their ions in water or acid.

The most widely used radioactive cow in nuclear medicine is $^{99}MO$–$^{99m}Tc$ cow. The $^{99}Mo$ produced at the reactor or cyclotron is embedded permanently in an Ion exchange resin. The resin containing the $^{99}Mo$ (half-life 67 days), which is constantly transforming into $^{99m}Tc$ (half-life 6 hours), is then shipped to the hospital. When needed, the physician washes the $^{99m}Tc$ from the cow for injection into the patient. The $^{99}Mo$ does not wash of the resin.

**a$^{137}Cs$-$^{137m}Ba$ cow will be milked to produce $^{137m}Ba$**

This cow is made by embedding $^{137}$Cs on an ion exchange resin. In the cow, the reaction

$$^{137}_{55}\text{Cs} - \,^{137m}_{56}\text{Ba} + e^-$$

occurs where the $e^-$ is an electron. When the cow is "milked," the Ba–137m is obtained and the reaction

$$^{137m}_{56}\text{Ba} - \,^{137}_{56}\text{Ba} + \gamma\text{(gamma radiation)}$$

occurs. Using a counter, the gamma radiation is measured. This gives the number of atoms of barium-137m transforming per second (the counts/second measured by the counter). The counts/second at various times will be determined over a period of several minutes. From the data obtained, a plot of the radiation detected as a function of the elapsed time will be made. The half-life of the milked $^{137m}$Ba is less than three minutes. Therefore, one must work quickly to measure the decrease in transformations taking place in the Ba-137m.

Below is a typical graph of counts per minute versus time. *The half-life is the time where the counts/minute is one-half its original value.*

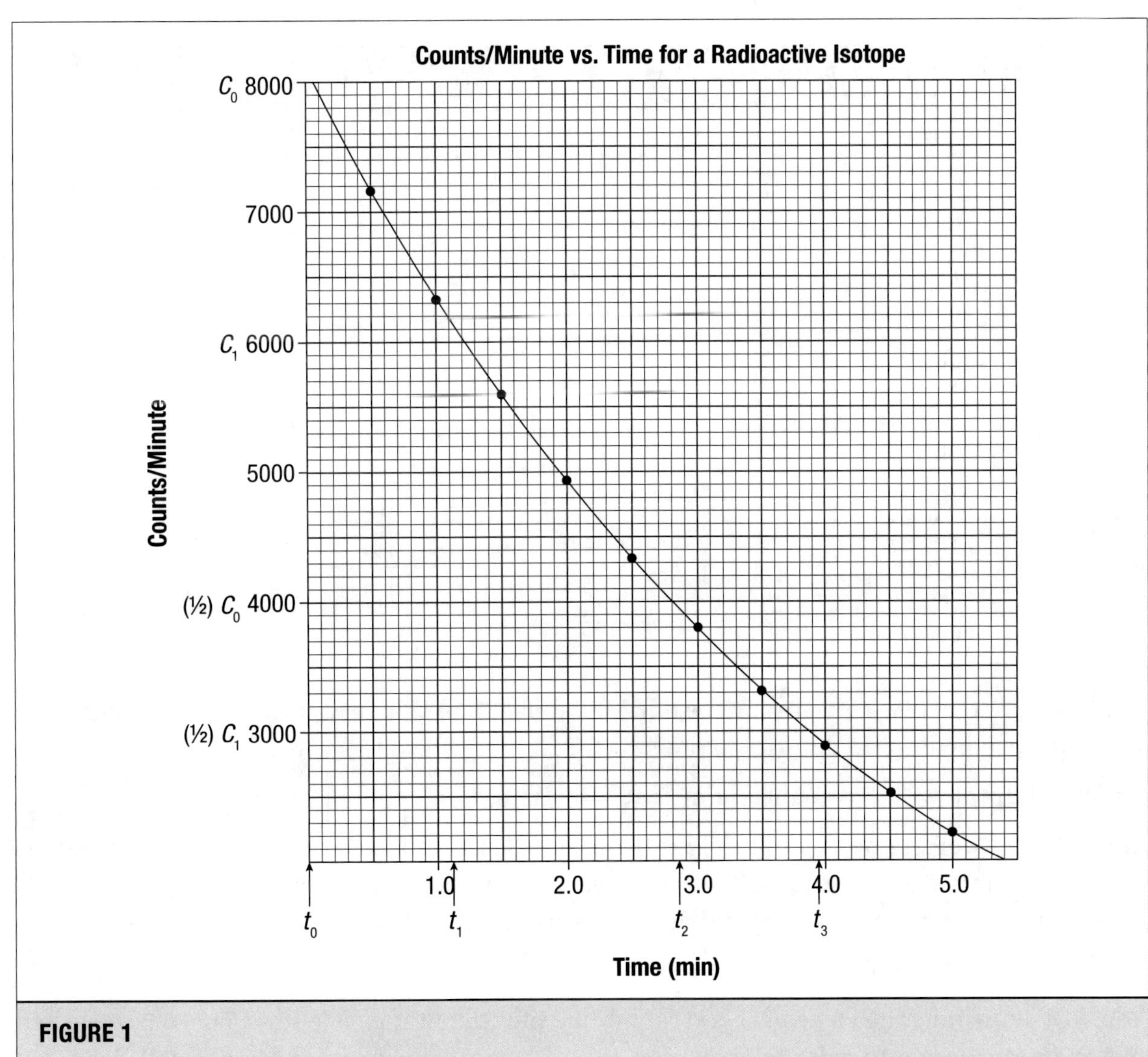


**FIGURE 1**

The following counts/minute and *corresponding* times are read from the graph as marked on the graph.

| | |
|---|---|
| $C_0$ = 8000 counts/minute | $t_0$ = 0 minutes (starting time) |
| $C_1$ = 6000 counts/minute | $t_1$ = 1.12 minutes |
| (½) $C_0$ = 4000 counts/minute | $t_2$ = 2.8 minutes |
| (½) $C_1$ = 3000 counts/minute | $t_3$ = 3.94 minutes |

The half-life is the time required for the counts/minute to drop to one-half its value (*starting anywhere*).

**time to go from 8000 to 4000 counts/min is $t_2 - t_0$ = 2.8 − 0 = 2.8**

**time to go from 6000 to 3000 counts/min is $t_3 - t_1$ = 3.94 − 1.12 = 2.82 minutes**

These values are the same to within the accuracy of the graph. The half-life is then 2.8 minutes. Note that it does not matter where one starts; the time for the counts per minute to become half its value is the same.

**TABLE 1**

| Time (sec) | Time (min) | Counts/Minute |
|---|---|---|
| 15 | 0 | 1095 |
| 30 | 0.5 | 999 |
| 60 | 1.0 | 886 |
| 90 | 1.5 | 755 |
| 120 | 2.0 | 686 |
| 150 | 2.5 | 580 |
| 180 | 3.0 | 495 |
| 210 | 3.5 | 452 |
| 240 | 4.0 | 376 |
| 270 | 4.5 | 357 |
| 300 | 5.0 | 278 |

You will construct two graphs for the data set above. The goal is to have two different plots that can each be used to determine the half-life, $t_{½}$, for this data set.

Your first graph will plot counts/minute versus time (min).

You will then determine three half-lives from this plot, just like the example on the previous page (and similar to Figure 16.12 on page 711 of Silberberg and Amateis). Be sure to use a ruler when you draw in the perpendicular "drops" to the x-axis.

Your second graph will plot ln(counts/minute) versus time(min).

You will determine the slope of this straight line plot to obtain a k value. (There is a worked out example for you to refer to shown on page 1086 of Silberberg and Amateis.) Then you will calculate half-life by using the relationship $t_{½}$ = 0.693/$k$ (Equation 16.7 on page 711).

# DATA SHEET

Name: _______________________________     Grade: _______________________________

Date Experiment Performed: _______________     Days Late: _______________________

CRN of Lab Section: _______________________     Instructor's Initials: _______________

## Questions and Problems

1. Write the symbols for the following types of radiation

   **a.** Alpha particle: ____________          **c.** Gamma ray: ____________

   **b.** Beta particle: ____________

2. List some shielding materials adequate for protection from

   **a.** Alpha particles:

   **b.** Beta particles:

   **c.** Gamma rays:

3. Complete the nuclear equations by filling in the correct symbols:

   **a.** $^{27}_{13}\text{Al} +$ ____________ $\rightarrow {}^{24}_{11}\text{Na} + {}^{4}_{2}\text{He}$          **c.** $^{96}_{40}\text{Zr} +$ ____________ $\rightarrow {}^{1}_{0}\text{n} + {}^{99}_{42}\text{Mo}$

   **b.** $^{131}_{53}\text{I} \rightarrow {}^{0}_{-1}\text{e} +$ ____________

4. The bacteria in some foods are sterilized by placing the food near a source of ionizing radiation. Does that mean that the food becomes radioactive? Explain.

5. Write a balanced nuclear equation for the alpha decay for each of the following radioactive isotopes:

   **a.** $^{208}_{84}\text{Po}$

   **b.** $^{232}_{90}\text{Th}$

   **c.** $^{251}_{102}\text{No}$

   **d.** $^{220}_{86}\text{Rn}$

6. Write a balanced nuclear equation for the beta decay for each of the following radioactive isotopes:

   **a.** $^{25}_{11}\text{Na}$

   **b.** $^{20}_{8}\text{O}$

   **c.** $^{92}_{38}\text{Sr}$

   **d.** $^{42}_{19}\text{K}$

7. Complete each of the following nuclear equations:

   **a.** $^{28}_{13}\text{Al} \rightarrow \underline{\hspace{2cm}} + ^{0}_{-1}\text{e}$

   **c.** $^{66}_{29}\text{Cu} \rightarrow ^{66}_{30}\text{Zn} + \underline{\hspace{2cm}}$

   **b.** $\underline{\hspace{2cm}} \rightarrow ^{86}_{36}\text{Kr} + ^{1}_{0}\text{n}$

   **d.** $\underline{\hspace{2cm}} \rightarrow ^{4}_{2}\text{He} + ^{234}_{90}\text{Th}$

8. Complete the following bombardment reactions:

   **a.** $^{9}_{4}\text{Be} + ^{1}_{0}\text{n} \rightarrow \underline{\hspace{2cm}}$

   **c.** $\underline{\hspace{2cm}} + ^{1}_{0}\text{n} \rightarrow ^{24}_{11}\text{Na} + ^{4}_{2}\text{He}$

   **b.** $^{32}_{16}\text{S} + \underline{\hspace{2cm}} \rightarrow ^{32}_{15}\text{P}$

9. Technetium-99m is an ideal radioisotope for scanning organs because it has a half-life of 6.0 hours and is pure gamma emitter. Suppose that 80.0 mg were prepared in the technetium generator this morning. How many milligrams would remain after

   **a.** one half-life: ________

   **c.** 18 hrs: ________

   **b.** two half-lives: ________

   **d.** 24 hrs: ________

# Discussion

Radioactivity occurs when a proton or neutron breaks down in the nucleus of an unstable atom, or the particles in the nucleus are rearranged. Then a particle or energy called **nuclear radiation** is emitted from the nucleus. The nucleus has undergone nuclear **decay.** The most typical types of radiation include alpha particles ($\alpha$), beta particles ($\beta$), and gamma rays ($\gamma$).

**alpha decay** $^{147}_{62}\text{Sm} \rightarrow ^{143}_{60}\text{Nd} + ^{4}_{2}\text{He}$

*alpha particle ($\alpha$)*

**beta decay** $^{40}_{20}\text{Ca} \rightarrow ^{40}_{21}\text{Sc} + ^{0}_{-1}\text{e}$

*beta particle ($\beta$)*

**gamma decay** $^{167m}_{68}\text{Er} \rightarrow ^{167}_{68}\text{Er} + ^{0}_{0}\gamma$

*gamma ray ($\gamma$)*

# Which Salts Make Good Hot and Cold Packs?



## Objectives

At the completion of the lab, you should be able to

- identify salts that can serve as reactants in chemical cold and hot packs;
- determine the heats of reactions for the different salts;
- recommend the best salts for use in hot or cold packs based on the heat-of-reaction data;
- identify any patterns between the heat behavior of different salts and salt mass; and
- provide recommendations regarding the mass of solid salt for use in cold and hot packs.

# 20-QUESTION PRE-LAB ASSIGNMENT

CRN:________________________________________     Your Name: ___________________________________

Date Submitted: _______________________________     TA's Name: ___________________________________

*Please write your answers legibly in the space provided. Some answers will actually consist of three or four pieces of information that are clearly written on the board during the conversation **or** clearly stated verbally during the presentation. You must include all aspects of the answer to receive full credit. These answers are due at the beginning of the lab meeting.*

**1.** What are we using our sense of touch to determine this week?

**2.** What is a **bond enthalpy,** and what units are used for bond enthalpies?

**3.** Neither positive nor negative signs are included with the bond enthalpy values in Table 9.3, on page 314 of the textbook. Why?

**4.** Which bonds are the weakest and have the smallest bond enthalpies? Which are the second weakest bonds and have larger bond enthalpies? Which are the strongest bonds with the largest bond enthalpy values? (Show this left to right below, using the carbon-to-carbon bonds as examples.)

**5.** In **CHE 111L,** the **dissolution** process consisted of two distinct steps? List them below.

Dissolution = _________________________________ + _________________________________

**6.** True or false? The dissolution process is an energy trade-off. _________________________________

**7.** Which **three steps** do we use to represent **dissolution** in **CHE 112L?**

Dissolution =

**8.** Show whether each of the three above steps requires making bonds or breaking bonds.

**9.** Show the energy consideration ("enthalpy word") and sign (+ or –) for each step in Question 7.

**10.** From whose perspective are the signs of the energy consideration written?

**11.** What is **the system** generally? What specifically is **the system** here?

**12.** Use the concept of endothermic versus exothermic process(es) to show how/why the test tube would feel cold or hot to our touch.

Dissolution =                            +                            +

Endo- or exothermic?                     +                            +

+ sign or – sign?                        ,                            ,

Group the above processes within parentheses (just like in the pre-lab video) to designate the conditions that give an endothermic salt and an exothermic salt.

**13.** How many times (trials) will you test one endothermic salt *and* one exothermic salt? How many grams will you use each time?

**14.** What does **approximately accurately** mean? (You can relate it to this lab.)

**15.** Fill in the table below with the temperature change for 2 g, for 4 g, and then **per 1 g of salt.**

| TABLE 1.1 | | |
| --- | --- | --- |
| **grams of Salt** | **$\Delta T$ (°C)** | **Normalize = $\Delta T$ per 1 gram of Salt (°C/g)** |
| 2.02 | | |
| 3.97 | | |
| **Average (°C/g)** | | |

**16.** What is the name of the device that we use to study constant pressure calorimetry? And what is the important relationship between $\Delta H$ and $q?$

**17.** Although we did not hold the test tube that contained the salt in the sample calculation, based on the temperature data collected, we *know* that **this is an exothermic salt.** How?

**18.** Define **specific heat.** Then list the specific heat for water *and* for very dilute solutions.

**19.** How do we get to a mass of **101 g** for use in the calorimetry equation?

**20.** Show the progression of units from $q \rightarrow \Delta H \rightarrow$ **molar $\Delta H$.**

# BACKGROUND

You have been hired as a consultant to a company that manufactures cold and hot packs, which are often used to treat an injury such as to a muscle or shoulder. The chemical reaction in the cold pack absorbs heat, while the chemical reaction in the hot pack releases heat.

The company recently received bequests of salts (chlorides, nitrates, and sulfates). Your job is to identify salts for use as reactants and provide guidelines regarding the mass of solid salt to be used in the hot or cold packs.

## Teams

You will work in pairs for this lab. You and your partner are to investigate the merits of different salts for potential use in cold or hot packs.

**TABLE 1.2  Salt Assignments**

| Even-Numbered Stations | | Odd-Numbered Stations | |
|---|---|---|---|
| Name | Formula | Name | Formula |
| Ammonium Nitrate | $NH_4NO_3$ | Ammonium Chloride | $NH_4Cl$ |
| Calcium Chloride | $CaCl_2$ | Magnesium Chloride | $MgCl_2$ |
| Magnesium Sulfate | $MgSO_4$ | Sodium Nitrate | $NaNO_3$ |
| Potassium Chloride | $KCl$ | Sodium Chloride | $NaCl$ |
| Potassium Nitrate | $KNO_3$ | | |

<table>
<tr><td>SAFETY!<br>WASTE<br>DISPOSAL</td><td>All waste materials may be disposed down<br>the sink with copious amounts of water.</td></tr>
</table>

## Sample Calculations for an Exothermic Solution

A team recorded the following data for the addition of $MgSO_4$ to 100 mL of water.

### Trial 1

2.02 g resulted in a temperature increase of 7.4°C.

### Trial 2

3.97 g resulted in a temperature increase of 15.2°C.

**TABLE 1.3**

|  | **Mass** | **$\Delta T$** | **$\Delta T$/Mass of Salt** |
|---|---|---|---|
| Trial 1 | 2.02 g | 7.4°C | 7.4°C/2.02 g = 3.66°C/g |
| Trial 2 | 3.97 g | 15.2°C | 15.2°C/3.97 g = 3.83°C/g |
| **Average** | 3.7°C/g | | |

So on average, the addition of 1 g of $MgSO_4$ to 100 mL water (we may assume 100 g water) will cause a temperature increase of 3.7°C.

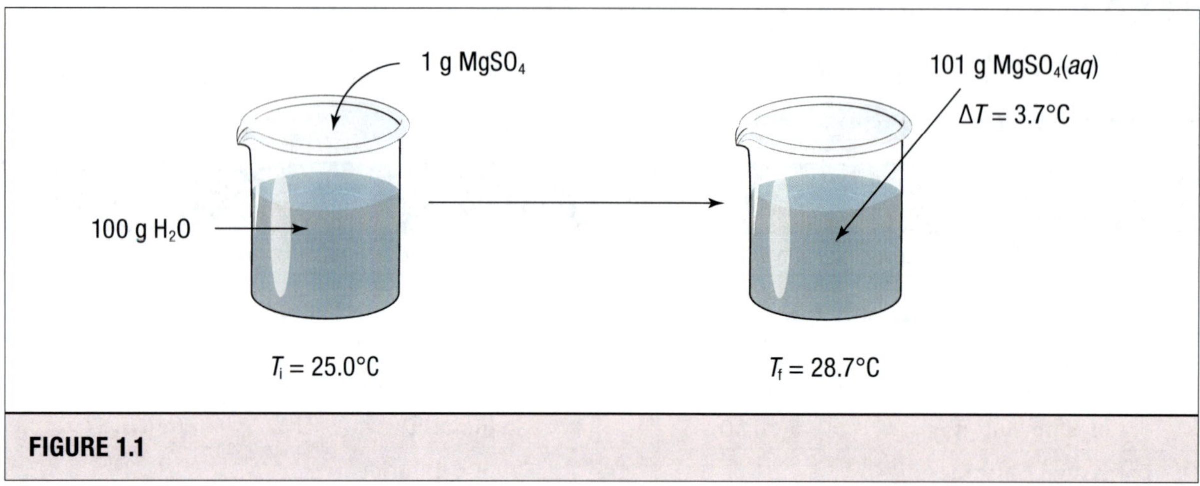


**FIGURE 1.1**

How do we use this data to estimate the amount of heat released? (See Table 1.8 on the Data Sheet.)

> **$q$(heat) = $\Delta H$** (only because we are measuring changes under the condition of constant pressure)
>
> **$q$ = (mass of solution) × (specific heat of the solution) × $\Delta T$**
>
> **$q$ = 101 g × (4.184 J)/(g · °C) × 3.7°C**
>
> **$q$ = 1.6 kJ** (Note the conversion to kJ.)

**That is, the heat release by dissolving 1 g of $MgSO_4$ in water is 1.6 kJ.**

# PROCEDURE

## Part I: Qualitative Heat Studies

Your preliminary studies are qualitative tests to determine whether heat is liberated or absorbed when a salt is dissolved in water. ***Be safe! Use a test tube holder; some salt solutions may generate large amounts of heat.***

1. Place about 10 mL of distilled water into a test tube, and use a spatula to transfer about 1 cm$^3$ or about 1 g of the assigned team salt to the test tube. Mix the contents with a stirring rod, and cautiously touch the test tube to the palm of your hand to see whether it becomes warmer, cooler, or neither. Repeat this procedure for each assigned salt. Refer to the information below, and enter your observations in Table 1.4 for each assigned salt.

   - For "Temperature Change," enter "Increase," "Decrease," or "Neither."

   - For "Type of Reaction," enter "Endothermic" if the mixture becomes cooler, "Exothermic" if the mixture becomes warmer, or "unsure."

   - For the "Sign of $q$," enter "+" if heat is absorbed by the mixture, "−" if heat is liberated by the mixture, and "±" if you detect no heat change.

| TABLE 1.4 Qualitative Observations | | | |
|---|---|---|---|
| **Salt** | **Temperature Change** | **Type of Reaction** | **Sign of $q$** |
|  |  |  |  |
|  |  |  |  |
|  |  |  |  |
|  |  |  |  |
|  |  |  |  |

2. Compare your results with any other team investigating the same salts. Repeat any team tests if there are discrepancies in the results. When you are done, collect the results of all teams by entering the results of the different teams in Table 1.5.

3. On the basis of the compiled results, select and record the three best potential salt candidates for hot packs and the three best potential salt candidates for cold packs.

## Compiled Team Data

| TABLE 1.5  Compiled Qualitative Observations | | | |
|---|---|---|---|
| Salt | Temperature Change | Type of Reaction | Sign of $q$ |
| $NH_4Cl$ | | | |
| $NH_4NO_3$ | | | |
| $CaCl_2$ | | | |
| $MgCl_2$ | | | |
| $Mg(NO_3)_2$ | | | |
| $MgSO_4$ | | | |
| KCl | | | |
| $KNO_3$ | | | |
| NaCl | | | |
| $NaNO_3$ | | | |

The three best salts for hot packs:

The three best salts for cold packs:

# Part II: Quantitative Heat Studies

Based on the observations in Part I, teams along with the TA are to have discussed and chosen the three best candidates for cold packs and the three best candidates for hot packs for further quantitative investigations. Each team will then conduct two quantitative heat investigations, one using a hot pack candidate and one using a cold pack candidate. For comparison purposes, each salt should be tested by a minimum of two teams.

Hot pack salt candidate:

Cold pack salt candidate:

4. For heat measurements, obtain two (rather than one) concentric Styrofoam cups to minimize heat exchange with the surroundings. Use two concentric Styrofoam cups with a temperature probe and a magnetic stirrer inserted into the cup through a cardboard or plastic lid. This apparatus serves as a constant pressure calorimeter.

5. Using a graduated cylinder, transfer 100.0 mL of distilled water into the calorimeter. Set up a ring stand and clamp to suspend a stirring rod and 0.1°C thermometer or temperature probe passing through the lid into the water in the calorimeter. Make sure that the temperature probe does not touch the bottom or the sides of the calorimeter.

6. Weigh ~2 grams of your assigned hot pack candidate salt, and report the mass to the nearest 0.01 g by entering the mass in Table 1.6.

7. While stirring the water, record the temperature of water for about 4–5 minutes until it has come to thermal equilibrium with the temperature probe. Record this initial temperature in Table 1.6. Add the weighed salt to the water with vigorous mixing. Observe the temperature of the solution every 30 seconds for about 8–10 minutes until the temperature has stabilized and achieved thermal equilibrium. Record this final temperature value in Table 1.6.

8. Repeat Steps 1–3 using 4-gram samples of the hot pack candidate salt and record your data in Table 1.6. Calculate $\Delta T/g$ for each test result and an average $\Delta T/g$. Does there appear to be a pattern with regard to the amount of salt tested versus the temperature change?

9. Repeat the entire procedure using the cold pack candidate.

## Team Data

| TABLE 1.6  Mass and Temperature Data per 100 mL Water | | | | | | |
| --- | --- | --- | --- | --- | --- | --- |
| **Exothermic Salt** | | | | **Endothermic Salt** | | |
| Mass (g) | | | | Mass (g) | | |
| Initial $T$°C | | | | Initial $T$°C | | |
| Final $T$°C | | | | Final $T$°C | | |
| $\Delta T$ (°C) | | | | $\Delta T$ (°C) | | |
| $\Delta T/$Mass (°C/g) | | | | $\Delta T/$Mass (°C/g) | | |
| Avg. $\Delta T/$Mass (°C/g) | | | | Avg. $\Delta T/$Mass (°C/g) | | |

10. Do you observe a pattern with regard to the mass or the tested salt and the observed temperature change? Refer to the data to offer an explanation for your conclusion.

11. Collect the different team results for the average $\Delta T$/mass for the different salts using Table 1.7 below.

| TABLE 1.7 Compiled Data, Average $\Delta T$/m for Exothermic and Endothermic Salts | | | | | |
|---|---|---|---|---|---|
| Exothermic Salt | Team | Average $\Delta T$/Mass (°C/g) | Endothermic Salt | Team | Average $\Delta T$/Mass (°C/g) |
| | | | | | |
| | | | | | |
| | | | | | |
| | | | | | |
| | | | | | |
| | | | | | |
| | | | | | |
| | | | | | |
| | | | | | |
| | | | | | |
| | | | | | |
| | | | | | |

# DATA SHEET

Name: _______________________________     Grade: _______________________________

Date Experiment Performed: _______________________     Days Late: _______________________________

CRN of Lab Section: _______________________     Instructor's Initials: _______________________________

| General Grading Items | 20 Points |
|---|---|
|  |  |
|  |  |
| 20-Question Pre-Lab Assignment | /20 |
|  |  |
|  |  |
| **Total** | **/20** |

| Data Analysis and Interpretation for Part I | 40 Points |
|---|---|
| Question 1 | /10 |
| Questions 2a and 3a | /15 |
| Questions 2b and 3b | /15 |
| **Total** | **/40** |

| Data Analysis and Interpretation for Part II | 40 Points |
|---|---|
| Table 1 | /20 |
| Questions 1 and 2 | /10 |
| Question 3 | /10 |
| **Total** | **/40** |

# Data Analysis and Interpretations for Part I

## Information

- When a salt dissolves in water, opposing heat processes are involved. An input of heat energy is required to separate the associated ions in the solid salt, and an input of heat energy is required to separate the solvent molecules; heat energy is released when the water molecules surround (solvate) the ions. Whether heat is liberated or absorbed **overall** depends upon the sum of the energies associated with these two processes.

- The reaction of dissolving a salt in water (dissolution reaction) is represented below. MA represents the salt, where M is any metal cation and A is the non-metal anion:

$$MA(s) \xrightarrow{\ H_2O(l)\ } M^+(aq) + A^-(aq)$$

**1.** Write the complete dissolution (dissociation) reactions for all tested salts below:

$NH_4Cl(s) \rightarrow$

$NH_4NO_3(s) \rightarrow$

$CaCl_2(s) \rightarrow$

$MgCl_2(s) \rightarrow$

$Mg(NO_3)_2(s) \rightarrow$

$MgSO_4(s) \rightarrow$

$KCl(s) \rightarrow$

$KNO_3(s) \rightarrow$

$NaCl(s) \rightarrow$

$NaNO_3(s) \rightarrow$

**2.** This question deals with salts that are cold pack candidates.

    **a.** List all salts yielding endothermic reactions that are candidates for cold packs.

**b.** Based on the data, which process involves more energy in the endothermic reactions? Specifically, is more energy involved in separating the ions or solvating the ions? Please explain.

**3.** This question deals with salts that are hot pack candidates.

**a.** List all salts yielding exothermic reactions that are candidates for hot packs.

**b.** Based on the data, which process involves more energy in the exothermic reactions? Specifically, is more energy involved in separating the ions or solvating the ions? Please explain.

# Data Analysis and Interpretations for Part II

Based on the compiled class data, complete Table 1.8 indicating the heat absorbed or liberated and the $\Delta H$ (heat per mole) for each salt. Calculate the heat absorbed or liberated based on the compiled average $\Delta T$/mass.

## Information

$$q = m \times SH \times \Delta T$$

where $m$ is the mass of the solution in grams, $SH$ is the specific heat of the solution, $SH = 4.184 \ J/g \cdot °C$ *and* $q_p = \Delta H$. (The enthalpy change is equal to the heat transferred at constant pressure; here, the constant pressure is atmospheric pressure.)

| TABLE 1.8 Heat Absorbed or Liberated for Each Salt | | | | | | | |
|---|---|---|---|---|---|---|---|
| Exothermic Salt | Average $\Delta T$/Mass (°C/g) | Heat Released (q) | $\Delta H$ (kJ/mol) | Endothermic Salt | Average $\Delta T$/Mass (°C/g) | Heat Absorbed (q) | $\Delta H$ (kJ/mol) |
| | | | | | | | |
| | | | | | | | |
| | | | | | | | |

Calculations:

4. Put the three endothermic salts in order from maximum to minimum potential as candidates for use in chemical cold packs based on the comparative values for $\Delta T$/mass, amount of heat absorbed, and $\Delta H$.

   a. Endothermic salt ranking based on comparative $\Delta T$/mass ratios.

                      >                    >

   b. Endothermic salt ranking based on comparative amount of heat absorbed.

                      >                    >

   c. Endothermic salt ranking based on comparative $\Delta H$ (heat per mole of salt).

                      >                    >

   d. Team ranking or candidates for use in chemical cold packs. Justify your final selection and rating of the candidates.

5. Put the three exothermic salts in order from maximum to minimum potential as candidates for use in chemical cold packs based on the comparative values for $\Delta T$/mass, amount of heat released, and $\Delta H$.

   a. Exothermic salt ranking based on comparative $\Delta T$/mass ratios.

                      >                    >

   b. Exothermic salt ranking based on comparative amount or heat absorbed.

                      >                    >

   c. Exothermic salt ranking based on comparative $\Delta H$ (heat per mole of salt).

                      >                    >

   d. Team ranking or candidates for use in chemical hot packs. Justify your final selection and rating of the candidates.

6. Based on your data, what observable patterns, if any, are there in the heat behavior of chloride versus nitrate salts, and +1 (monovalent) cation versus +2 cation (divalent) salts?

# Colligative Properties
## Freezing Point Depression

## Objectives

At the completion of the lab, you should be able to

- empirically determine the freezing point depression constant, $K_f$, for water; and

- identify an unknown, electrolyte compound by determining the van't Hoff factor, $i$.

# 20-QUESTION PRE-LAB ASSIGNMENT

CRN:_______________________________________     Your Name: _______________________________________

Date Submitted: _______________________________     TA's Name: _______________________________________

*Please write your answers legibly in the space provided. Some answers will actually consist of three or four pieces of information that are clearly written on the board during the conversation **or** clearly stated verbally during the presentation. You must include all aspects of the answer to receive full credit. These answers are due at the beginning of the lab meeting.*

**1.** What kind of solution can be treated as if it were "pure water"? (We did this in Experiment 1 last week.)

**2.** Can we treat the solutions in lab today as if they are "pure water"?

**3.** What are the four colligative properties?

**4.** How are the names of these properties indicative? (In other words, what happens to the freezing point, boiling point, and vapor pressure of a solution compared to those properties of the pure solvent?)

**5.** What is the root meaning of the word "colligative"?

**6.** What is the "common thread" in all four colligative properties?

**7.** How is this "common thread" proportional to the colligative properties?

**8.** In general, what do we compare when we study freezing point depression (or any other colligative property)?

**9.** What is a lattice? What special effect did our producer incorporate to help you visualize a lattice?

**10.** In pure ice, only water molecules are incorporated into the lattice. What gets included in the lattice of ice when a solution freezes?

**11.** What makes an aqueous solution "aqueous"?

**12.** What is the rule we learned that will help us to predict whether a solute will dissolve in water? Why don't oil and water mix?

**13.** What is the name of that convenient class of compounds that will dissolve in water? Draw its "family tree" and define the name.

**14.** Distinguish between **electrolyte** versus **non-electrolyte.**

**15.** What is **the** equation to use for FPD? Define all the terms *except i* and *m* in the space below.

**16.** Define molality, m. Define molarity, M. Why do we use molality rather than molarity for our study of freezing point depression?

17. What is $i$? How do we "calculate" a value for $i$? Show that "calculation" below for urea, sodium chloride, and magnesium chloride.

18. Prove that urea is expected to be soluble in water by showing its net permanent dipole moment.

19. Draw a sketch to show why the empirical $i$ values are usually not equal to the "calculated" $i$ values for ionic compounds, **but** they are always **exactly** equal for polar covalent non-electrolytes? What **is** the numerical value of $i$ for every single polar covalent non-electrolyte?

20. For what three different substances do we measure values of $T_f$ today, and what is (are) our goal(s) by measuring these $T_f$ values?

# BACKGROUND

Colligative properties are physical properties of solutions that depend on the total number of solute particles, but not the kind of particles, dissolved in solution. Colligative properties include boiling point elevation, freezing point depression, increasing osmotic pressure, and vapor pressure lowering. A solution (a solute dissolved in a solvent) will always have a higher boiling point, a lower freezing point, a higher osmotic pressure, and a lower vapor pressure than that of the pure solvent. The degree to which the colligative properties of a solution are affected is directly related to the number (concentration) of solute particles in the solution. Solutions that contain larger numbers of solute particles will have higher boiling points and lower freezing points than solutions that contain smaller numbers of solute particles. Since colligative properties depend only on the total number of solute particles present, the freezing point depression of a 0.1 m aqueous NaCl (a strong electrolyte) solution is 0.348°C, approximately twice that of a 0.1 m aqueous glucose (a non-electrolyte) solution, whose $T_f$ = 0.186°C.

## Teams

You will work in teams of two for this lab; however, students will turn in individual reports.

| SAFETY!<br>WASTE<br>DISPOSAL | **All waste materials from this experiment may be disposed down the sink with running water.** |
| --- | --- |

## Information

The freezing point of a solution is always lower than the freezing point of the pure solvent. The difference between the freezing point $T_f$ of a pure solvent and that of a solution is the freezing point depression, $\Delta T_f$:

$$\Delta T = iK_f m$$

For water $K_f$ is 1.86°C/m, m is the molality (mol solute/kg solvent), and *i* is the van't Hoff factor. The van't Hoff factor, *i,* indicates the extent of dissociation of strong electrolytes or the extent of ionization of weak electrolytes in aqueous solution. The van't Hoff factor is the ratio of the experimentally observed colligative property to the value for a non-electrolyte solution of the same concentration. For non-electrolyte solutes (e.g., urea, sucrose, etc.) the value of *i* is one. In infinitely dilute solutions in which no appreciable ion association occurs, the *i* value for a strong electrolyte such as NaCl should be two (NaCl dissociates into two ions, $Na^+$ and $Cl^-$); however, the *i* value for a 0.1 m aqueous NaCl solution is slightly less than two (*i* = 1.87) due to ion association in solution. (See Figure 13.27 on page 565 of the Silberberg and Amateis text.)

Since freezing can be difficult to discern by eye, a temperature versus time graph called a *cooling curve* is used instead.

# PROCEDURE

Connect a temperature probe to Channel 1 of the Vernier LabQuest2. Connect the power cord into the LabQuest2 and then into the outlet located nearest you. The power button is then located on the top left directly above the LabQuest2 logo. Both Channel 1 (CH 1) and the power button are shown in Figure 2.1. The stylus used to navigate the touch screen is then located on the right side of the system under the power cord shown in Figure 2.2.

**FIGURE 2.1**

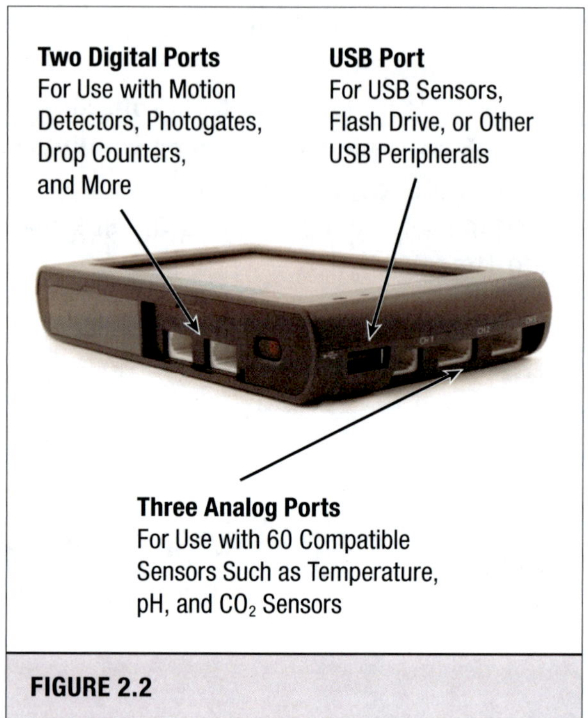


**FIGURE 2.2**

Once the LabQuest2 has powered on, locate the graph symbol at the top of the screen, and use the pen to navigate to this particular tab.

## Part I: Determining the Freezing Point for Water

1. Prepare an ice-salt-water bath by adding ~150 mL water to a 600-mL beaker. Add 20–25 g of salt and stir to dissolve (if using rock salt, there may be some undissolved crystals). Add ice to the beaker so that the interstitial spaces are filled completely with water, **and** the ice does not float off the bottom of the beaker. (More salt [or ice] may be added to the ice bath during the experiment; if the ice bath temperature is not sufficiently cold to freeze the solution, ask your lab instructor.)

2. Assemble the wire stirrer/temperature probe assembly shown in Figure 2.3. *Make certain the temperature probe passes through the wire loop.* Secure the assembly to a ring stand using a utility clamp if necessary.

3. Add approximately 40–45 mL of distilled water to a large test tube.

4. Insert the temperature probe/wire stirrer assembly into the ice bath water. Carefully monitor the temperature of the ice bath because once it has dropped below 2°C, data collection will need to begin. Click ▶ to begin collection.

5. Insert the test tube/temperature probe/stirrer assembly into the ice-salt-water bath. Be sure all of the liquid in the test tube is below the level of the ice-salt-water bath as shown in Figure 2.4.

6. Stir the water by moving the wire stirrer up and down in the test tube until the water is frozen. Super cooling ($T$ of the liquid drops below $T_f$ without freezing) will quite possibly occur, but upon freezing the temperature will rise and then remain constant as all the water becomes frozen.

7. Once the water is frozen and temperature has remained constant for 40–60 seconds, press stop ■.

8. The freezing temperature can be determined by finding the mean temperature in the portion of the graph with nearly constant temperature.

9. At the top of the screen click the word "Analyze." A drop box will appear. Then click on "Statistics." A small box will then appear beside it that says "Temperature." This will give you all important statistics needed to complete the Data Sheet.

10. It is important to then save this file. At the top of the screen click "file." A drop box will once again appear. Click "Save" and name this file "Water."

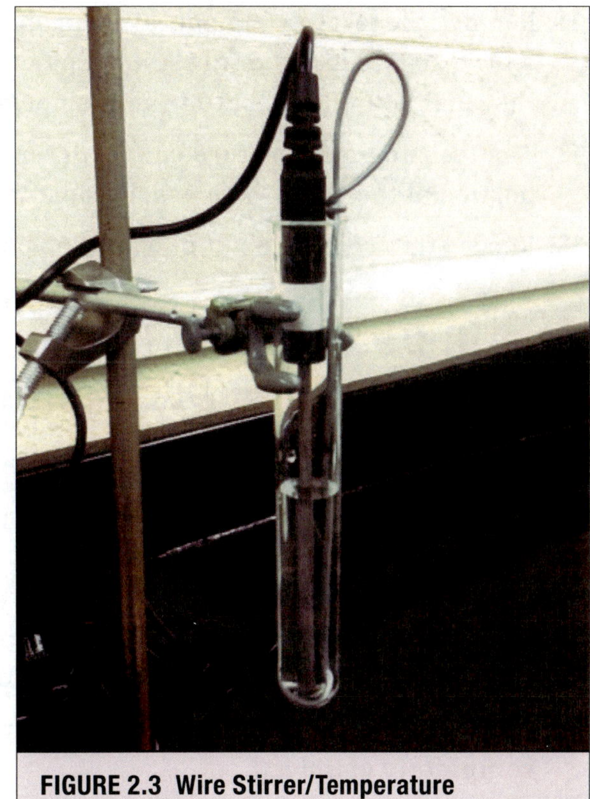

**FIGURE 2.3  Wire Stirrer/Temperature Probe Assembly**

**FIGURE 2.4  Wire Stirrer/Temperature Probe Assembly in Ice-Bath**

11. Remove the test tube/temperature probe/stirrer assembly from the ice–salt–water bath, and re-secure it to the ring stand. Remove the ice/water mixture from the test tube, and thoroughly dry the test tube and temperature probe/stirrer with a Kimwipe.

12. The freezing temperature can be determined by finding the mean temperature in the portion of the graph with nearly constant temperature.

13. Record the freezing point of water in your notebook.

## Part II: Determining the Freezing Point Depression Constant for Water

14. Place the large test tube in a 125-mL Erlenmeyer flask, and record the mass to the nearest 0.001 g in your notebook. (**Note:** The test tube must be dry.) Add approximately 15 grams of water to the test tube, and reweigh the flask/test tube assembly. Record the mass to the nearest 0.001 g in your notebook, and determine the mass of the water by difference.

15. Place a sheet of weighing paper on the balance, and weigh out ~1 gram of urea. Record the mass to the nearest 0.001 g in your notebook.

16. Use a disposable pipette to stir the solution in the test tube until the unknown solid is **completely** dissolved.

17. Insert the temperature probe/wire stirrer assembly into the water. Carefully monitor the temperature of the ice bath because once it has dropped below 2°C, data collection will need to begin. Click ▶ to begin collection.

18. Insert the test tube/temperature probe/stirrer assembly into the ice bath. Be sure all of the liquid in the test tube is below the level of the ice bath as shown in Figure 2.4.

19. Record the cooling curve, and determine the freezing point for the urea solution (see Part I, Steps 6–10).

20. Perform a second trial of the aqueous urea solution by repeating Part II, Steps 14–19.

## Part III: Determining the van't Hoff Factor of an Unknown Electrolyte

Obtain 15 mL of a 0.200 m solution containing an unknown, ionic compound as the solute and record the unknown number in your notebook.

21. Record the cooling curve, and determine the freezing point of the electrolytic solution. It may be necessary to drain some of the aqueous solution from the ice-salt-water-bath and add more salt or ice.

22. Perform a second trial using the same unknown solution.

**At the conclusion of this experiment, there should be five cooling curves, one of water and two each of the aqueous urea and unknown ionic solutions. Save all cooling curves to a flash drive.**

# DATA SHEET

Name: _______________________________   Grade: _______________________________

Date Experiment Performed: _______________   Days Late: _______________________

CRN of Lab Section: _____________________   Instructor's Initials: ___________________

| General Grading Items | 20 Points |
|---|---|
| Copies of Lab Pages Attached; Labeled with Name and Date, Complete Information, Readable, Data Recorded Matches Results Given in Report | |
| All Safety Rules Were Followed | |
| Waste Was Properly Disposed of and Lab Area Was Cleaned | |
| Evaluation of Student Performance Overall (Student Was on Time, Followed Safety Rules, Performed the Lab Correctly and Within the Time Allowed, Etc.) | |
| 20-Question Pre-Lab Assignment | /20 |
| **Total** | **/20** |

| Data Analysis and Interpretation for Parts II and III | 80 Points |
|---|---|
| Question 2 | /5 |
| Table 1, Including Sample Calculations | /20 |
| Questions 3 (graph, 5 points) and 4 (5 points) | /10 |
| Table 2, Including Sample Calculations | /20 |
| Questions 5 (graph, 5 points) and 6 (5 points) | /10 |
| Question 7 | /5 |
| Question 8 | /10 |
| **Total** | **/80** |

# Part II: $K_f$ of Water

1. The freezing point of distilled water as determined in Part I = 

2. The freezing point you measured was probably not exactly zero. Why might this be the case? In other words, what might cause an experimentally measured freezing point not be exactly 0.00°C for water?

**TABLE 2.1**

| Trial | Urea Solution | | | $\Delta T_f$/°C | $K_{f,expt}$/°C · m$^{-1}$ |
| --- | --- | --- | --- | --- | --- |
| | Mass of Urea/g | Mass of Water/kg | Molality | | |
| 1 | | | | | |
| 2 | | | | | |
| Average | | | | | |

3. Show a sample calculation for one of your $K_f$ values reported in Table 2.1. **Supply a landscaped cooling curve showing both aqueous urea trials.**

4. For water, the accepted value of $K_f$ is 1.86°C/m. Calculate the percent error of your experimental $K_f$.

# Part III: van't Hoff Factor of an Unknown Electrolyte

| TABLE 2.2 | | | | |
|---|---|---|---|---|
| **Trial** | **Unknown Electrolytic Solution** | | **$\Delta T_f$/°C** | **$i_{expt}$** |
| | **Unknown** | **Molality** | | |
| 1 | | | | |
| 2 | | | | |
| **Average** | | | | |

5. Show a sample calculation for one of your $i$ values reported in Table 2.2. **Supply a landscaped cooling curve showing both aqueous unknown trials.**

**6.** Using Table 2.3 below, identify your unknown electrolyte:

| TABLE 2.3 | | | |
| --- | --- | --- | --- |
| **Salt** | **van't Hoff $i$** | **Salt** | **van't Hoff $i$** |
| $MgCl_2$ | 2.70 | $K_2SO_4$ | 2.32 |
| $MgSO_4$ | 1.21 | KBr | 1.88 |

**7.** Suppose that in Part II, Step 3, a student did not completely dissolve the urea sample.

   **a.** Will the freezing point depression, $\Delta T_f$, of the solution be larger or smaller than it would have been if the compound was dissolved completely? Justify your answer with an explanation.

   **b.** Will the calculated freezing point depression constant, $K_f$, be larger or smaller than it would have been if the compound was dissolved completely? Justify your answer with an explanation.

**8.** In order to avoid the possibility of 2-hour delays this winter, EKU will be stockpiling de-icing salts to help keep the roads and sidewalks clear. Facility Services has been quoted the following prices:

- $K_2SO_4$: \$5.35/kg
- $MgSO_4$: \$4.55/kg

Which salt provides a better value? Support your answer with a calculation.

# Kinetics I
## The Decolorization of Brilliant Blue Dye

## Objectives

This experiment is intended to show how the reaction rate is experimentally determined. At the completion of the lab, you should be able to

- write an expression for the overall reaction rate;
- determine the order of a reaction using the method of "average" rates; and
- graphically determine the "average" rate, rate constant, and half-life of a first order reaction.

# 20-QUESTION PRE-LAB ASSIGNMENT

CRN:_______________________________     Your Name: _______________________________

Date Submitted: _______________________     TA's Name: _______________________________

*Please write your answers legibly in the space provided. Some answers will actually consist of three or four pieces of information that are clearly written on the board during the conversation* **or** *clearly stated verbally during the presentation. You must include all aspects of the answer to receive full credit. These answers are due at the beginning of the lab meeting.*

1. What *type* of reaction is "the decolorization of brilliant blue dye"?

2. What are the first two of the Big 3 questions that chemical analyses attempt to answer?

3. What must particles do in order for reactions to occur?

4. Define "success" for kinetics by listing the two aspects of a successful collision.

5. What is our non-chemistry example for "success"?

6. What can the *rate expression* be written for?

**7.** Write out the *rate expression* for today's reaction of dye and bleach using all the reactants and products.

**8.** What is the "easiest" change for us to monitor in this chemical reaction? Why?

**9.** What are the units for rate(s)?

**10.** Which measurements do we use to calculate $\Delta t$ and $\Delta[\text{dye}]$?

**11.** What is the name of the useful method that we adapt for this experiment?

**12.** Outline the actual process we will follow in order to collect the appropriate data.

**13.** By collecting all these data and performing all these necessary calculations, what is it our goal to deduce?

**14.** How is the rate law unlike the rate expression?

**15.** Write out the general format of the rate law. Three variables (three "letters" in the general equation) must be empirically determined. What are these variables called?

**16.** During the experiment, which measurement do we hold constant for the first five trials, **and** what is that constant amount?

**17.** What are $R^*$ and $C$, and how do we use them to calculate $a$ and $b$?

**18.** What are the expected values for $a$ and $b$?

**19.** Draw a representation of each of the three plots that kinetics data lend themselves to $-[A]$ versus $t$, $\ln [A]$ versus $t$, and $1/[A]$ versus $t$. Of these three data plots, how many of them will be linear? What does a linear plot tell us about the kinetics aspect of our data?

**20.** How do the plots described in Question 19 above relate to radioactivity? (Draw out all three plots as part of your explanation.) Which one is linear? What does this tell us about radioactivity?

# BACKGROUND

The kinetics of a reaction involves two principal assessments:

1. The rate, or speed, at which a reaction takes place. The rate of a reaction is determined by measuring the changes in amounts (or concentrations) of reactants (or products) which occur during a reaction as a function of time.

2. The mechanism, or series of steps, by which a reaction takes place. The mechanism of a reaction is determined by making a detailed study of the changes in structure and composition which occur with the particles involved in a reaction as these particles convert from reactants to products. We will study the mechanism of this reaction in the next lab assignment.

## The Reaction Rate Expression

The chemical rate law for a chemical reaction, can generally be quantitatively expressed by recognizing that, at constant temperature, the initial rate of a reaction is proportional to the molar concentration product of the reactants with each concentration raised to a definite exponential power. Application of this statement to a reaction yields the general chemical rate expression for the reaction. To illustrate, consider the following hypothetical reaction.

$$2A + B \rightarrow \text{products}$$

Based on the above statement we can write

$$\text{rate} \propto [A]^m[B]^n$$

where the square brackets, [ ], mean molar concentration of the substance in the brackets.

For the exponents $m$ and $n$:

1. These exponents must be evaluated by experiment. They cannot be determined from the coefficients in the stoichiometric equation.

2. It is an experimental fact that the sum of all such exponents (overall order) in any rate expression ($m + n$, in this case) rarely exceeds 3. But there are some exceptions.

3. An individual exponent ($m$ or $n$) is commonly 1 or 2 and occasionally 3. An individual exponent is 0 (zero) if the rate is independent of the concentration of the corresponding reactant; and an individual exponent can even be fractional (such as ½). Fractional exponents frequently appear in rate expressions for complex reactions—which we do *not* study in CHE 112L.

## Reaction Order

Reaction order is the numerical relationship which exists between the reaction rate and the concentration terms in the rate law. For a reaction where $m$ and $n$ are each 1 (one), the rate of this reaction is said to be first order in $A$ (since $m$ is 1), first order in $B$ (since $n$ is 1), and second order overall—since $m + n$ equals 2. That is, the overall order of a reaction is simply the sum of the individual exponents on the concentration terms. The significance or reaction order is that it shows how reaction rate changes as a function of change-in-concentration for a particular reactant. For instance, consider the following rate law:

$$\text{rate} \propto [A]^m$$

If the reaction is first order in $A$ ($m = 1$; also first order overall) then doubling $[A]$ results in doubling the rate. But, if the reaction is second order in $A$ ($m = 2$; also second order overall) then doubling $[A]$ results in quadrupling the rate, i.e., $[A]^2$ becomes $[2A]^2$ when the concentration is doubled, thus increasing the rate by a factor of $2^2$ or 4.

$$\text{rate} \propto [2A]^2 = 2^2[A]^2 = 4[A]^2$$

# The Rate Constant

For any reaction studied at constant temperature, the rate divided by the corresponding concentration terms from the rate expression is a constant. This constant, which must be determined empirically, is called the "rate constant" and is designated as $k$. To illustrate

$$\frac{\text{rate}}{[A]^m[B]^n} = k$$

In other words, $k$ is a *proportionality constant.* Notice that the units of $k$ will be dependent on the overall order of the reaction. k units are compiled in Table 3.1.

# Determination of Rate Constant

For a reaction whose rate depends only on a single reactant $[A]$, the following table shows the relationships for overall reaction order, rate equation, reactant concentration $[A]$ in terms of time ($t$): units on rate constant, and half-life ($t_{1/2}$). Half-life is the time required for the initial concentration to decrease to one-half its initial value, where $[A]_0$ indicates initial concentration of reactant $A$.

**TABLE 3.1**

| Order | Rate Expression | $[A]$ in Terms of $t$ | Units of $k$ | $t_{1/2}$ |
|---|---|---|---|---|
| 0 | $k[A]^0 = k$ | $[A] = [A]_0 - kt$ | $\text{mol} \cdot L^{-1} \cdot s^{-1}$ | $\dfrac{[A]_0}{2k}$ |
| 1 | $k[A]^1 = k$ | $\ln [A] = \ln [A]_0 - kt$ | $s^{-1}$ | $\dfrac{\ln 2}{k}$ |
| 2 | $k[A]^2$ | $\dfrac{1}{[A]} = \dfrac{1}{[A]_0} + kt$ | $L \cdot mol^{-1} \cdot s^{-1}$ | $\dfrac{1}{k[A]_0}$ |

The "[A] in terms of time" expressions are particularly important because they can be used to determine $k$. For example, if the order is zero, a plot of $[A]$ versus $t$ is linear (i.e., of the form $y = mx + b$) with a slope of $-k$ and a $y$-intercept of $[A]_0$. Thus, $k$ can be determined if $[A]$ can be measured as a function of $t$. Similarly, if the order is one, a plot of $\ln [A]$ versus $t$ is linear with a slope of $-k$ and a $y$-intercept of $\ln [A]_0$.

The "$t_{1/2}$" expressions are also important because if $t_{1/2}$ can be determined, a check on the value for $k$ can be obtained. The simplest of such cases is illustrated by a first order reaction because $t_{1/2} = \ln 2 / k$ does not depend on $[A]_0$ (or on $[A]$) and is therefore constant. The following graph illustrates this:

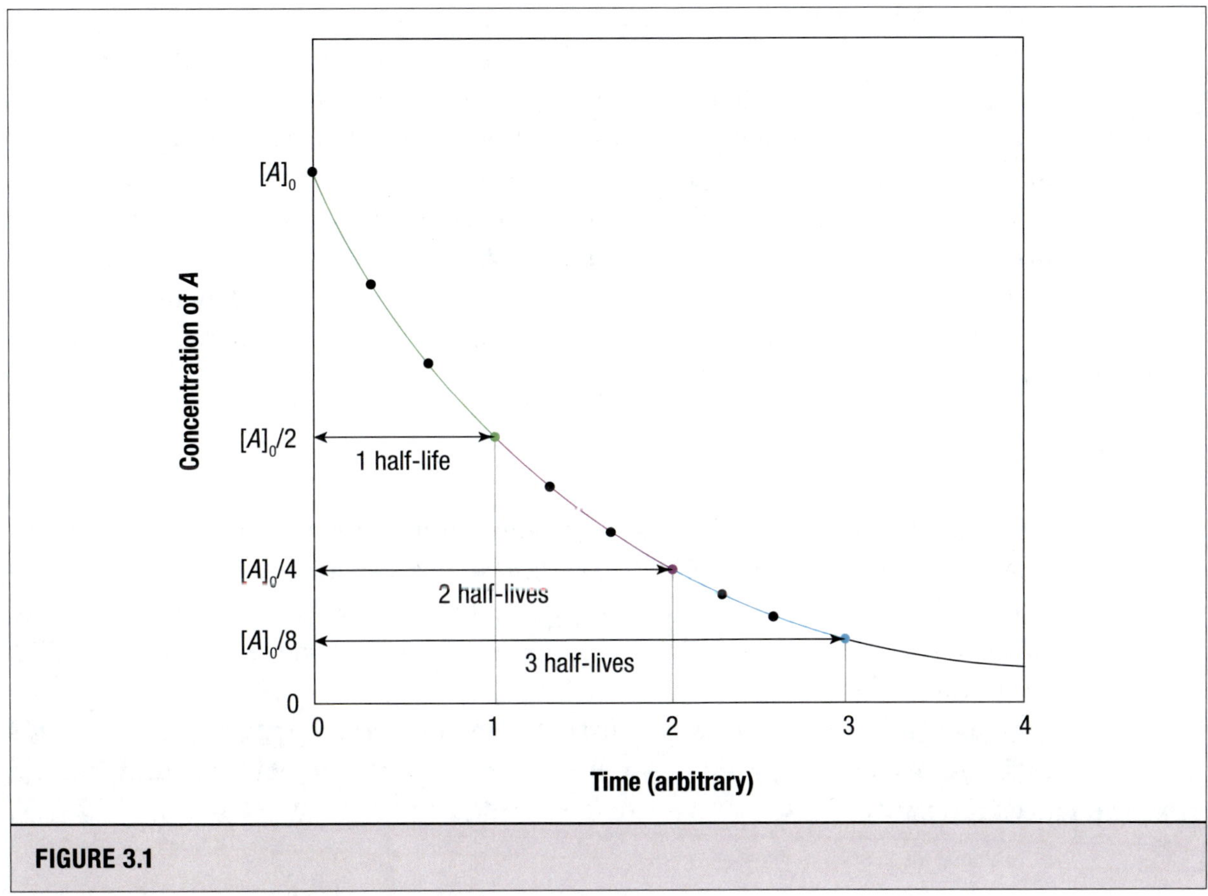


**FIGURE 3.1**

# The Reaction to Be Studied

The reaction to be studied in this assignment is the decolorization of blue food dye by the bleaching action of common household bleach. Depending on initial concentrations of bleach and dye, solution acidity, and solution temperature, the total time required for the decolorizing reaction is such that reasonably reliable kinetic data can be obtained by observation of dye-color disappearance as a function of time.

Bleach is principally an alkaline aqueous solution of sodium hypochlorite, NaClO. So it can be inferred that the hypochlorite ion, ClO⁻, is the bleaching (oxidizing) agent in bleach. But, kinetic investigations on the chemistry of bleach indicate that the actual bleaching agent is probably hypochlorous acid. We will investigate this in a subsequent lab. Nevertheless, the overall stoichiometric equation for the bleach/dye reaction may be written as

$$\textbf{dye}(aq) + \textbf{ClO}^-(aq) \rightarrow \textbf{o-dye}(aq) + \textbf{Cl}^-(aq)$$

where o-dye is the oxidized dye product. The hypochlorite ion is acting as an oxidizing agent.

The progress of the bleach/dye reaction will be monitored exclusively by observation of dye-color disappearance. Furthermore, the bleach/dye reaction is not reversible and will always go to completion (i.e., the dye color will always disappear completely). Therefore, each run of this reaction will involve measuring the time required for the dye color to disappear. This will first be measured as a function of initial bleach concentration and thereafter as a function of initial dye concentration. Now note that for simplification, actual concentrations of bleach and dye will be ignored and "relative rates" rather than actual rates will be determined for each run. So, since reaction progress is to be monitored by dye-color disappearance, and since rates for reactions are studied by measuring changes ($\Delta$) in molar concentration of reactant or product per corresponding change in time, we will use

$$\textbf{rate} = \frac{\Delta\textbf{[dye]}}{\Delta\textbf{time}}$$

Finally, note that the results which you will obtain should show that this reaction is first order in bleach as well as first order in dye, and therefore second order overall. **But,** this experiment has been designed such that for any "run" [bleach] is very large relative to [dye]. (This condition ensures that the dye color will always disappear completely in a fairly short period of time.) Therefore, for each run made, [bleach] will remain essentially constant as [dye] goes to zero. This means that from an **overall order** standpoint, this reaction will behave as if it is first order. Be sure to keep this important point in mind as you do experiment.

# PROCEDURE

**N.B.,** When doing this experiment be sure to have a timepiece which will allow you to measure time in seconds.

## Part I: The Nature of Reactants

The ability of common bleach to decolorize different dyes will be observed. In each of two small, clean Erlenmeyer flasks, prepare 20.0 mL of dilute bleach by mixing 5.0 mL of stock bleach with 15.0 mL of distilled water, each poured from a graduated cylinder. Be careful with bleach as it can burn! Clean up any bleach spills carefully! (**Note:** Measure all liquid volumes carefully throughout this experiment.) Prepare to measure the time to the nearest second required to decolorize both crystal violet dye and brilliant-blue (BB) food dye. Each will decolorize totally.

Add 2 drops of crystal violet dye to one of the flasks, swirl thoroughly, and immediately begin to time. (**Note:** Repeated swirling will not be necessary during the bleaching process.) **In your notebook,** record the total time required for the dye color to fade to a fairly constant, almost colorless appearance similar to the bleach solution itself. Repeat this investigation by adding a drop of BB dye to the second flask.

Retain the solution of the decolorized BB dye-product as a color standard for marking the complete bleaching of BB dye in the experiments to follow.

**In your notebook,** comment on which decolorization investigated above appears to be easier to study from the standpoint of reliable time measurement? Why?

## Part II: Reaction Rate and Reactant Concentration, Reaction Order, and Rate Equation

The order of the reaction with respect to each reactant will be determined by studying the effect of change in reactant concentration on reaction rate. This will allow determination of the rate equation.

### Order with Respect to Bleach

In separate labeled flasks, prepare the bleach solutions listed in the following table. Note that each solution has a total volume of 20.0 mL. (Use a graduated cylinder for these volume measurements!)

**TABLE 3.2**

| Run | Volume (mL) | |
| --- | --- | --- |
| | Stock Bleach | Distilled Water |
| 1 | 2.0 | 18.0 |
| 2 | 4.0 | 16.0 |
| 3 | 5.0 | 15.0 |
| 4 | 8.0 | 12.0 |
| 5 | 10.0 | 10.0 |

Measure the temperature of one of the solutions, noting that this will be the $T$ for each solution provided that the stock bleach and distilled water are at the same $T$. Check this before proceeding! **In your notebook,** record $T$ to the nearest 0.1°C.

Treat each solution with one drop of BB dye and immediately swirl. (Note that all runs can be performed simultaneously if you are careful!) Measure and record **in your notebook** the time required for each solution to decolorize and achieve a color similar to that of the bleached BB dye-solution prepared in Part I. Repeat all runs which are within 10 seconds of each other. Repeat any run which results in uncertain time measurement.

Calculate a "relative rate" for each satisfactory run, and enter this information in Table 3.4. (Show all calculations in your notebook.) Note that relative rates are calculated as $1000/\Delta t$ in order to get numbers (rates) which are larger than one. Let [dye] equal the number of drops of BB dye used in the run. (Remember that you do not know any actual dye or bleach concentrations.)

The general rate equation for the bleach/dye reaction is given by

$$\text{rate} = k[\text{bleach}]^a[\text{dye}]^b = k[\text{OCl}^-]^a[\text{dye}]^b$$

where the most probable values for $a$ and $b$ are 0, 1, or 2. For the data just collected, $k$ and [dye] are constants. By comparing rates and bleach concentrations for various runs, the value of "$a$" can be determined. Complete Table 3.5 to serve this purpose.

## Order with Respect to Dye

The value for $b$ will now be determined by a series of runs in which the dye concentration is varied while the bleach concentration is held constant.

To each of three separate, labeled flasks, add 10.0 mL distilled water and the corresponding number or drops of BB dye for Runs 6 through 8. Swirl (see Table 3.3). Treat each (on a one-at-a-time basis, or simultaneously, if you are careful) with 10.0 mL stock bleach and immediately reswirl. Measure and record **in your notebook** the time required for each solution to decolorize and achieve a color similar to that of the bleached BB dye-solution prepared in Part I. Repeat any run resulting in uncertain time measurement. Note that the elapsed time for Run 9 has been determined above in Run 5. You should perform Run 10 to test the precision of your data by comparing the time elapsed for Run 10 to the time elapsed for Run 5.

**TABLE 3.3**

| Run | Drops of Dye | Volume (mL) | |
| --- | --- | --- | --- |
| | | Stock Bleach | Distilled Water |
| 6 | 2 | 10.0 | 10.0 |
| 7 | 3 | 10.0 | 10.0 |
| 8 | 4 | 10.0 | 10.0 |
| 9 | 1 | 10.0 | 10.0 |

Calculate a relative rate for each satisfactory run and enter in the Table 3.6. For the data just collected, $k$ and $[ClO^-]$ are constants. By comparing rates and dye concentrations for various runs, the value of "$b$" can be determined. Complete Table 3.7 to serve this purpose, again show all calculations in your notebook.

# Part III: Determination of Rate Constant and Confirmation that Overall Reaction Order Appears to be One

A simple spectrometric method will be used to determine the reaction rate constant $k$ and to confirm apparent overall reaction order. This method depends on the fact that colored materials (such as BB dye) absorb light in the visible region of the spectrum. The absorbing ability or "absorbance" of the dye solution is directly proportional to the concentration of the dye in the solution. Since the BB dye used in this experiment absorbs at a maximum of 635 nm (nanometers), spectrometric readings of absorbance taken at 635 nm can be used to monitor the change in dye concentration which occurs as the bleaching process proceeds. The data just obtained will be used to calculate $k$ for the bleach/dye reaction from a $t_{1/2}$ plot and from a ln [reactant] versus time plot. Correspondence between these $k$s will confirm that the overall reaction order for this reaction appears to be 1. That is, as it is being studied, this reaction should behave as if it is an overall first order reaction.

See Appendix B concerning operation of the colorimeter.

Refer to the **Directions for Using the Colorimeter with LabQuest,** in a separate file in BlackBoard.

For **Step 5** in those directions, ***do NOT mix crystal violet and NaOH. Do*** mix bleach and brilliant blue (BB) dye as described below.

5. In a small beaker, mix 2.0 mL stock bleach with 18.0 mL distilled water. Add 2 drops of BB dye and swirl thoroughly. Fill the cuvette ¾ full with this solution, and immediately place the cuvette in the colorimeter and click ▶ Collect . The absorbance reading will rapidly approach 1.00 as decolorization occurs in the cuvette. Logger Pro will begin recording absorbance versus time as soon as the absorbance drops below ~0.7 absorbance units.

Save your data. It will be used to prepare properly labeled and scaled graphs of Absorbance vs. Time, Natural Log of Absorbance vs. Time, and Inverse Absorbance vs. Time. These graphs will be turned in with the Data Sheet.

# Directions for Using the Colorimeter with LabQuest

1. Prepare a *blank* by filling a cuvette ¾ full with distilled water. To correctly use cuvettes, remember to

   - wipe the outside of each cuvette with a lint-free tissue;
   - handle cuvettes only by the top edge of the ribbed sides;
   - dislodge any bubbles by gently tapping the cuvette on a hard surface; and
   - always position the cuvette so the light passes through the clear sides.

## Colorimeter Users Only (spectrometer users proceed to the spectrometer section)

2. Connect the colorimeter to LabQuest and choose "New" from the "File" menu.

3. Calibrate the colorimeter.

   a. Place the blank in the cuvette slot of the colorimeter and close the lid.

   b. Press the "<" or ">" buttons on the colorimeter to set the wavelength to 635 nm (red). Then calibrate by pressing the "CAL" button on the colorimeter. When the LED stops flashing, the calibration is complete.

4. Set up the data-collection mode.

   a. On the "Meter" screen, tap "Rate." Change the data-collection rate to 1 sample/ second.

   b. Change the data-collection length to 200 seconds. Select "OK."

   c. Proceed to Step 5.

## Both Colorimeter and Spectrometer Users

5. *Do this quickly!* To initiate the reaction, simultaneously pour the 10-mL portions of crystal violet and sodium hydroxide into a 250-mL beaker, and stir the reaction mixture with a stirring rod. Empty the water from the cuvette. Rinse the cuvette twice with ~1-mL amounts of the reaction mixture, fill it ¾ full, and place it in the device (colorimeter or spectrometer). Close the lid on the colorimeter. Start data collection.

6. Absorbance data will be collected for 200 seconds. You may stop data collection early if desired. Discard the beaker and cuvette contents as directed by your instructor.

7. Analyze the data graphically to decide if the reaction is zero, first, or second order with respect to crystal violet.

   - Zero order: If the current graph of absorbance versus time is linear, the reaction is *zero order.*

   - First order: To see if the reaction is first order, it is necessary to plot a graph of the natural logarithm (ln) of absorbance versus time. If this plot is linear, the reaction is *first order.*

   - Second order: To see if the reaction is second order, plot a graph of the reciprocal of absorbance versus time. If this plot is linear, the reaction is *second order.*

8. Follow these directions to create a calculated column, ln absorbance, and then plot a graph of ln absorbance versus time.

   a. Tap the "Table" tab to display the data table.

   b. Choose "New Calculated Column" from the "Table" menu.

   c. Enter the name (ln Abs) and leave the units field blank. Select the equation, Aln(X). Use absorbance as the column for X, and 1 as the value for A. Select "OK."

   d. A graph of ln absorbance versus time will be displayed.

9. Follow these directions to create a calculated column, 1/Absorbance, and then plot a graph of 1/Absorbance versus time:

   a. Tap the "Table" tab.

   b. Choose "New Calculated Column" from the "Table" menu.

   c. Enter "1/Absorbance" as the name, and leave the units field blank. Select the euqtion, A/X. Use "Absorbance" as the column for X, and 1 as the value for A. Select "OK."

   d. A graph of 1/Absorbance versus time will be displayed.

10. To see any of the three plots again:

    a. Tap the vertical-axis label of the graph.

    b. Choose "Absorbance," "ln Absorbance," or "1/Absorbance."

11. Optional: Print a copy of the table.

12. Optional: Print a copy of the graph that was most linear.

## Processing the Data

**13.** Was the reaction zero, first, or second order, with respect to the concentration of crystal violet? Explain.

**14.** Calculate the rate constant, $k$, using the *slope* of the linear regression line for your linear curve ($k = -$slope for zero and first order and $k =$ slope for second order). Be sure to include correct units for the rate constant. **Note:** This constant is sometimes referred to as the *pseudo rate constant* because it does not take into account the effect of the other reactant, OH⁻.

**15.** Write the correct rate law expression for the reaction, in terms of crystal violet (omit OH⁻).

**16.** Using the printed data table, estimate the half-life of the reaction; select two points, one with absorbance value that is about half of the other absorbance value. The *time* it takes the absorbance (or concentration) to be halved is known the *half-life* for the reaction. (As an alternative, you may choose to calculate the half-life from the rate constant, $k$, using the appropriate concentration-time formula.)

# DATA SHEET

Name: _______________________________  Grade: _______________________________

Date Experiment Performed: _______________________  Days Late: _______________________________

CRN of Lab Section: _______________________  Instructor's Initials: _______________________________

| General Grading Items | 20 Points |
|---|---|
| 20-Question Pre-Lab Assignment | /20 |
|  |  |
|  |  |
|  |  |
| **Total** | **/20** |

| Data Analysis and Interpretation for Order with Respect to Dye and Bleach | 40 Points |
|---|---|
| Data Table | /20 |
| Order with Respect to Both Dye and Bleach | /20 |
| **Total** | **/40** |

| Data Analysis and Interpretation for Determination of Rate Constant | 40 Points |
|---|---|
| Reasonable Absorbance vs. Time Data | /5 |
| Question 1 | /15 |
| Question 2 | /15 |
| Question 3 | /5 |
| **Total** | **/40** |

# Order with Respect to Bleach

**TABLE 3.4**

| Run | Volume (mL) | | Time (s) | Relative Rate ($s^{-1}$) $\left(\dfrac{1000\,\Delta[\text{dye}]}{\Delta t}\right)$ | Relative Bleach Concentration |
| --- | --- | --- | --- | --- | --- |
| | Stock Bleach | Distilled Water | $\Delta t$ | | |
| 1 | 2.0 | 18.0 | | | 1.0 |
| 2 | 4.0 | 16.0 | | | 2.0 |
| 3 | 5.0 | 15.0 | | | 2.5 |
| 4 | 8.0 | 12.0 | | | 4.0 |
| 5 | 10.0 | 10.0 | | | 5.0 |

**TABLE 3.5**

| Run Ratio | Relative Rate Ratio ($R^*$) | log R | Relative Bleach Concentration Ratio (C) | log C | Calculated $a$ $\log R^*/\log C$ |
| --- | --- | --- | --- | --- | --- |
| 2/1 | | | 2.0/1.0 | | |
| 3/2 | | | 2.5/2.0 | | |
| 3/1 | | | 2.5/1.0 | | |
| 4/3 | | | 4.0/2.5 | | |
| 4/2 | | | 4.0/2.0 | | |
| 5/2 | | | 5.0/2.0 | | |

In your notebook, calculate the average value of "*a*."

Assuming the nearest integer value to be correct, *a* = 

# Order with Respect to Dye

**TABLE 3.6**

| Run | Volume (mL) | | | Time (s) | Relative Rate ($s^{-1}$) $\left(\dfrac{1000\,\Delta[\text{dye}]}{\Delta t}\right)$ | Relative Dye Concentration |
| --- | --- | --- | --- | --- | --- | --- |
| | Stock Bleach | Distilled Water | Drops of Dye | $\Delta t$ | | |
| 6 | 10.0 | 10.0 | 2 | | | 2.0 |
| 7 | 10.0 | 10.0 | 3 | | | 3.0 |
| 8 | 10.0 | 10.0 | 4 | | | 4.0 |
| 5 | 10.0 | 10.0 | 1 | | | 1.0 |

| TABLE 3.7 | | | | | |
|---|---|---|---|---|---|
| Run Ratio | Relative Rate Ratio ($R^*$) | log R | Relative Bleach Concentration Ratio (C) | log C | Calculated $b$ $\log R^*/\log C$ |
| 6/5 | | | 2.0/1.0 | | |
| 7/6 | | | 3.0/2.0 | | |
| 8/7 | | | 4.0/3.0 | | |
| 8/5 | | | 4.0/1.0 | | |

In your notebook, calculate the average value of "$b$."

Assuming the nearest integer value to be correct, $b$ =

Based on your experimental results, write the complete rate equation for this reaction.

# Determination of Rate Constant and Confirmation that Overall Reaction Order Appears to be One

1. Prepare properly labeled and scaled graphs of Absorbance vs. Time, Natural Log of Absorbance vs. Time, and Inverse Absorbance vs. Time.

2. Determine the half for the bleach/dye reaction from the Absorbance vs. Time graph. Read and report at least five half-lives from this plot and calculate an average $t_{1/2}$. Does your graph show that the bleach/dye reaction behaved like an overall first-order reaction? Use the half-life to calculate the rate constant.

**3.** Referring to the Determination of Rate Constant section, graphically determine $k$ by the appropriate graph from Question 1. Briefly describe how you determined $k$. Do not forget to include units!

# Kinetics II
## Temperature Dependence of the Decolorization of Brilliant Blue Dye

## Objectives

This experiment is intended to show how the temperature affects reaction rate. At the completion of the lab you should be able to

- interpret reaction energy diagrams;
- define activation energy and recognize the Arrhenius equation;
- determine the activation energy of a reaction using graphical methods; and
- use the Arrhenius equation to predict the rate constants of a reaction at different temperatures.

# 20-QUESTION PRE-LAB ASSIGNMENT

CRN:_________________________________________     Your Name:________________________________________

Date Submitted:_____________________________     TA's Name:_________________________________________

*Please write your answers legibly in the space provided. Some answers will actually consist of three or four pieces of information that are clearly written on the board during the conversation **or** clearly stated verbally during the presentation. You must include all aspects of the answer to receive full credit. These answers are due at the beginning of the lab meeting.*

**1.** What is our goal to find out with this experiment?

**2.** What are the Big 3 questions? ...

**3.** ... ***And*** the Big 3 answers from last time?

**4.** Our goal last time was the rate law. Write it down and reveal the overall order of the reaction.

**5.** What are the four factors that affect rate, ***and*** what is their effect on $k$?

**6.** What is the general rule of thumb for how changing the temperature affects the rate between, roughly, 0°C and 40°C?

**7.** What is the equation that helps us relate temperature and $k$ numerically? Name it and write it out.

**8.** What is $R$ in this equation? Include numerical values and units.

**9.** Define $E_a$.

**10.** How do we "linearize" the Arrhenius equation, and what is the resulting equation?

**11.** What is $A$ in the above equation, and what do we do with it in CHE 112L?

**12.** What is the example problem from your text book that has data similar to those you collect and use in lab today? (It was on pages 486–487 in the 6th edition, but on pages 489–490 in the 7th edition that you use.)

**13.** Show how we can obtain $E_a$ graphically from the Arrhenius plot.

**14.** Going back to the "non-linearized" form of the Arrhenius equation, show how $k$ is related to activation energy and temperature.

**15.** Draw a potential energy diagram for an oxidation reaction like we saw back in Chapter 6.

**16.** Fill in the blank: *All* oxidations are ________________ and relate the potential energy of the reactants to the potential energy of the products in an oxidation.

**17.** Redraw the PE diagram for an oxidation reaction (like you drew in Question 15 above), but this time include (and label) the activation energy.

**18.** Draw a PE diagram that shows the effect of a catalyst.

**19.** What exactly does the catalyst do?

**20.** What else should you do before you leave today?

# BACKGROUND

There are four main factors which influence reaction rate:

1. Concentration of the reactants

2. Surface area

3. Temperature or the reaction system

4. Catalysts

In the previous experiment we explored how the reaction rate is experimentally measured, how the concentrations of the reactants affect reaction rate, and how the rate law for the reaction is obtained.

## Reaction Energy Diagrams

Generally, the kinetic progress of a reaction can be conveniently represented by constructing an energy (potential energy, actually) diagram for the reaction. These diagrams express the energy of a reaction as reactants are transformed into products. To illustrate, consider the following hypothetical reaction and its reaction energy diagram. This diagram indicates there is an "energy barrier" of 125 kJ/mol.

This barrier is the amount of energy that the reactants must overcome in order to be con-

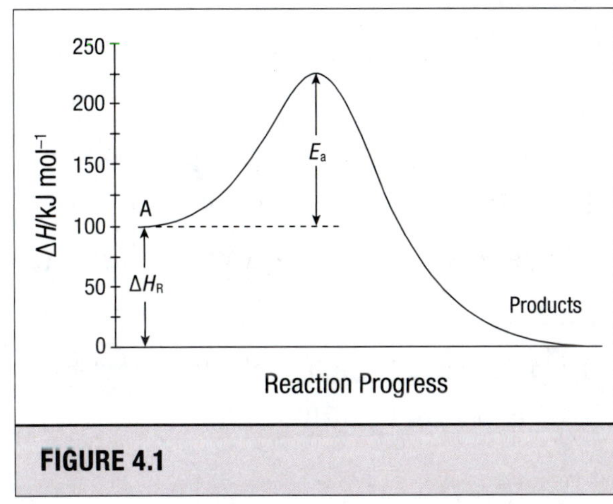


**FIGURE 4.1**

verted to products. This is known as the activation energy, represented as $E_a$. The diagram also indicates that the products are 100 kJ/mol lower in energy than the reactants. This is energy difference represents the change in enthalpy of the reaction. This hypothetical reaction has an activation energy of 125 kJ/mol, and it is exothermic with $\Delta H = -100$ kJ/mol.

## Temperature and Reaction Rate

Svante Arrhenius noted that the influence of temperature on the rate of many reactions can be predicted by the following equation:

$$\text{rate} \propto k \text{ and } k = Ae^{-E_a/RT}$$

where

- $k$ is the observed (experimentally determined) rate constant;

- $A$ is a factor which is proportional to the frequency of reactant-particle collisions as well as the probability that these collisions occur with appropriate in-space particle orientation so that reaction can occur;

- $e$ (2.7183) is the base of natural logarithms;
- $E_a$ is the observed activation energy ($E_a$ is typically reported in units of kJ/mol);
- $R$ is the ideal gas constant. $R = 8.314 \text{ J} \cdot \text{mol}^{-1} \cdot \text{K}^{-1}$; and
- $T$ is the absolute temperature in units of K.

By taking the natural logarithm of the Arrhenius equation, this converts the above expression into a linear equation ($y = mx + b$).

$$\ln k = -\frac{E_a}{R} \times \frac{1}{T} + \ln A$$

Thus if $ks$ (or rates) can be determined at two (or more) different temperatures, then the activation energies for the reaction can be obtained either algebraically or graphically, since the $\ln A$ term can be eliminated from the resulting simultaneous equations. That is, a graph of $\ln k$ or $\ln$ rate versus inverse temperature will be linear and have a slope of $-E_a/R$.

## The Reaction to Be Studied

The reaction to be studied in this assignment is the decolorization of blue food dye by the bleaching action of common, household bleach. The overall stoichiometric equation for the bleach/dye reaction may be written as

$$\text{dye}(aq) + \text{OCl}^-(aq) \rightarrow \text{dye} - \text{O}(aq) + \text{Cl}^-(aq)$$

where dye-O is the oxidized dye product. The hypochlorite ion is acting as an oxidizing agent.

The progress of the bleach/dye reaction will be monitored exclusively by observation of dye-color disappearance.

# PROCEDURE

**Note:** In doing this experiment, be sure to have a watch that will allow time measurement in seconds.

## Determination of Activation Energy

*For this experiment you will need just one Erlenmeyer flask filled with approximately 20 mL of the bleach solution and one 400-mL beaker filled with mostly ice and a little bit of water (so that your Erlenmeyer flask will not tip over).*

Note the laboratory supply of ice-cold bleach solution. This bleach solution should be at a $T$ of about 0°C. This bleach solution is 50% stock bleach/water.

1. Pour approximately 20 mL of the ice-cold bleach into a small, clean Erlenmeyer flask. Place the flask with the bleach in your 400-mL beaker, and put a little ice water (prepare as necessary from the supply or ice that is available in the laboratory) in the beaker to keep the bleach solution cold.

2. Prepare a table in you notebook with columns: "Initial Solution $T$ (°C)," "Final Solution $T$ (°C)," and "$\Delta t$(s)."

3. Measure the temperature (to the nearest 0.1°C) of the bleach solution, and record it in initial temperature column.

4. Add 1 drop of BB dye to the cold bleach solution, and immediately swirl to mix (only once or twice, not continuously). Swirl all later trials the same way—just once or twice, not continuously. Measure and record the time required for decolorization. As soon as decolorization is achieved, again measure and record the final solution temperature in the table.

5. Now allow the bleach/dye solution to warm a few degrees; when this solution is about 4–6°C, return this solution in its flask to the 400-mL beaker and again charge the beaker with some cold water to keep the bleach solution cold. Measure and record the initial solution temperature. Add another drop of BB dye, swirl, and repeat the above decolorization investigation. Remember to measure and record final solution temperature.

6. Again allow the bleach/dye solution to warm; this time to about 8–12°C, and repeat the decolorization investigation. Keep in mind that you may still use the same bleach/dye solution since [bleach] will practically not change during the reaction. But, if necessary, you may always obtain more of the original, ice-cold (0°C) bleach solution from the laboratory supply.

7. Continue by repeating the decolorization investigations at about 14–18°C and at about 18–22°C.

8. Conclude your measurements by doing one or two decolorization investigations at *T*s above room *T*. Do this by using hot water tap in the lab. The highest temperature run will use hot water directly from the tap. You will also do a decolorization investigation at a temperature between room temperature and the temperature of the hot water from the tap. Be sure to measure and record initial and final temperatures of the bleach/dye solutions.

9. As time permits, repeat any decolorization measurements which you feel did not produce reasonable or reliable results.

This data will be used to calculate the activation energy ($E_a$) for this reaction by graphic treatment of the Arrhenius equation.

# DATA SHEET

Name: _______________________________     Grade: _______________________________

Date Experiment Performed: _______________________     Days Late: _______________________________

CRN of Lab Section: _______________________     Instructor's Initials: _______________________________

| General Grading Items | 20 Points |
|---|---|
| 20-Question Pre-Lab Assignment | /20 |
| Copies of Lab Pages Attached; Labeled with Name and Date, Complete Information, Readable, Data Recorded Matches Results Given in Report | |
| Waste Was Properly Disposed of and Lab Area Was Cleaned | |
| Evaluation of Student Performance Overall (Student Was on Time, Followed Safety Rules, Performed the Lab Correctly and Within the Time Allowed, Etc.) | |
| **Total** | **/20** |

| Data Analysis and Interpretation | 80 Points |
|---|---|
| Data Table | /28 |
| Question 2, Arrhenius Plot | /20 |
| Question 3, Description | /17 |
| Question 3, Calculations | /15 |
| **Total** | **/80** |

**1.** Complete the following table.

**TABLE 4.1**

| Average Solution $T$* | | Time | Relative Rate $(s^{-1})$ | Natural Log Relative Rate | Inverse Temperature |
|---|---|---|---|---|---|
| $T$ (°C) | $T$ (K) | $\Delta t$ (s) | $\left(\dfrac{1000\ \Delta[\text{dye}]}{\Delta t}\right)$ | $\ln R$ | $1/T$ (K$^{-1}$) |
| | | | | | |
| | | | | | |
| | | | | | |
| | | | | | |
| | | | | | |
| | | | | | |
| | | | | | |
| | | | | | |
| | | | | | |

*The average temperature of the run is the average of the initial and final temperatures.

**2.** Prepare a properly labeled and scaled graph of Natural Log of Relative Rate vs. Inverse Temperature.

**3.** Using grammatically correct sentences, describe how this graph is used to determine the activation energy of the decolorization reaction. Report the activation energy in kJ/mol, and show any calculations used.

# The Hydrolysis of Ethyl Acetate

## Objective

At the completion of the lab, you should be able to

- determine the value of the equilibrium constant for the reaction of ethyl acetate and water to form acetic acid and ethyl alcohol.

## Reference

Silberberg and Amateis, *Chemistry, 8e*, pages 747–755.

# PRE-EXPERIMENT

## Objective

At the completion of the lab, you should be able to (eventually) determine the value of the equilibrium constant for the reaction of ethyl acetate and water to form acetic acid and ethyl alcohol.

## Pre-Procedure

To prepare the chemical systems needed for this experiment, eight vials must be set up at least one week in advance *exactly* as described below. Be sure to use the most accurate glassware available to you—and your best lab techniques—when you dispense these aliquots.

| TABLE 5.1  Volumes of Each Reagent Added to the Vials | | | | | |
|---|---|---|---|---|---|
|  | **3 M HCl** | **Ethyl Acetate** | **Water** | **Ethyl Alcohol** | **Acetic Acid** |
| Density of Compound → | 1.055 g/mL | 0.894 g/mL | 1.00 g/mL | 0.791 g/mL | 1.05 g/mL |
| Vials 1 & 2 | 5 mL | 0 | 0 | 0 | 0 |
| Vials 3 & 4 | 5 mL | 3 mL | 2 mL | 0 | 0 |
| Vials 5 & 6 | 5 mL | 0 | 0 | 4 mL | 1 mL |
| Vials 7 & 8 | 5 mL | 4 mL | 0 | 1 mL | 0 |

Once each vial is filled with reagents as indicated, it should be labeled (a small square of scotch tape that you can write on with ink or pencil is adequate), capped tightly, shaken thoroughly, allowed to sit for at least 20 minutes, and then shaken one more time before storage. Students should shake the vials at least once more before the actual laboratory period.

At the actual lab period, the vials will be titrated with NaOH (titration is simply the double replacement reaction between an acid and a base in stoichiometric amounts) to determine how much acetic acid is present in each vial. Each vial of the pair (1 & 2, 3 & 4, 5 & 6, 7 & 8) will be titrated separately, and then the average titration volume for each pair will be used to continue with calculations designed to determine the value of $K_{eq}$.

This experiment studies one of the three most important topics in chemistry—equilibrium!

# 20-QUESTION PRE-LAB ASSIGNMENT

CRN:_________________________________ Your Name:_________________________________

Date Submitted: _________________________ TA's Name: _________________________________

*Please write your answers legibly in the space provided. Some answers will actually consist of three or four pieces of information that are clearly written on the board during the conversation **or** clearly stated verbally during the presentation. You must include all aspects of the answer to receive full credit. These answers are due at the beginning of the lab meeting.*

**1.** What are we doing "this week" in lab?

**2.** What do the "$K$" and the "$C$" stand for?

**3.** What do we ***always*** need to remember when writing an equilibrium constant expression?

**4.** What is the name of this reaction? Write the equilibrium equation for this reaction with "reactants" (on the left) and "products" (on the right) using words and "shorthand."

**5.** Write the $K_C$ expression for this reaction ***as written.***

**6.** All of the reaction vials contain 10 mL of liquid in them, and so we can write $K_c$ without having to include volume to give ...

**7.** What do we do with the vial contents?

**8.** What concentration of NaOH will you be using in your lab experiment? What concentration of NaOH is used in the sample calculations?

**9.** What do we call Vials 1 & 2 (the vials with only HCl in them)?

**10.** Write out the titration equation, and show the dimensional analysis that ends with mmoles of acid and base.

**11.** Show the multiplication step we must do to obtain mmoles of base.

**12.** And then to obtain the mmoles of HCl.

**13.** What else is in the control vials besides HCl?

**14.** Show the sample calculations provided that allowed us to determine the mass of water in the controls.

**mass of the 5 mL aliquot = mass of HCl + mass of water**

**15.** The results to the above calculation remind us that every vial—all eight of them—contains how much HCl and how much water?

**16.** Show the calculation(s) done to obtain the number of mL of 1.24 M NaOH used to titrate the acetic acid in Vials 3 & 4.

**17.** Write out the I-C-E table for Vials 3 & 4.

**18.** Show the calculations for the mmoles of EtOAC and water that were initially present in Vials 3 & 4.

**19.** Finally, calculate $K_c$ for Vials 3 & 4.

**20.** What is the temperature dependence of $K_c$? And what does $K_c$ tell us about the relative ratio of products and reactants?

# BACKGROUND

## Discussion

For the general chemical equation aA + bB $\leftrightarrows$ cC + dD, the mass action expression (MAE) is written

$$K = \frac{[\text{products}]}{[\text{reactants}]} = \frac{[C]^c[D]^d}{[A]^a[B]^b}$$

The numerator involves the concentrations of the products and the denominator involves the concentrations of the reactants, each raised to a power equal to its coefficient in the balanced chemical equation.

Chemical reactions all react in one direction: toward equilibrium. Chemical equilibria are dynamic equilibria. Dynamic equilibrium is a condition in which macroscopically (large scale) there is no visible nor perceptible change with time in the state of the mixture of the components. Further, the concentrations of all components remain constant with time. Sub-microscopically (small scale, on the order of the size of molecules), there is a great deal going on. Forward and reverse reactions are constantly occurring. Chemical bonds are continually breaking and re-forming as individual molecules of reactants and products interchange. For instance, for the reaction

$$\underset{\substack{\text{ethyl acetate} \\ \text{(EtOAc)} \\ \text{(MM = 88)}}}{CH_3CH_2OOCCH_3} \quad + \quad \underset{\substack{\\ \\ \text{(MM = 18)}}}{H_2O} \quad \underset{\text{catalyst}}{\overset{\text{HCl}}{\leftrightarrows}} \quad \underset{\substack{\text{ethyl alcohol} \\ \text{(EtOH)} \\ \text{(MM = 46)}}}{CH_3CH_2OH} \quad + \quad \underset{\substack{\text{acetic acid} \\ \text{(HOAc)} \\ \text{(MM = 60)}}}{CH_3COOH}$$

at equilibrium, ethyl acetate is reacting with water to form ethyl alcohol and acetic acid (the forward reaction) while ethyl alcohol and acetic acid are reacting to form ethyl acetate and water (the reverse reaction). This happens in any mixture of these components; but, when the system is at equilibrium, the **rate** at which the forward reaction occurs is exactly equal to the **rate** at which the reverse reaction occurs. Ethyl acetate and water are reacting to form ethyl alcohol and acetic acid at the same rate they are being formed from the reaction of ethyl alcohol with acetic acid.

At equilibrium, it is found that the mass action expression (MAE)

$$K = \frac{[\text{ethyl alcohol}]\,[\text{acetic acid}]}{[\text{ethyl acetate}]\,[\text{water}]}$$

is constant, and its value is called the equilibrium constant, $K$ (or $K_{eq}$). This condition is true for all reactions. $K_{eq}$ does not depend on the individual concentrations of the components of the reaction equilibrium; there are many **sets** of individual concentrations that satisfy the MAE above. $K_{eq}$ does depend on the temperature of the reaction mixture.

One way to determine the value of the equilibrium constant is to determine the equilibrium concentrations of all components in the equilibrium mixture. Fortunately, it is often possible to determine the concentration of one of the components experimentally and then use the stoichiometry of the reaction to calculate the concentrations of the other components. This is done below in an example.

It is important that a reaction be given sufficient time to reach equilibrium. For the ethyl acetate and water reaction, even though a catalyst (HCl) is present, the mixture will be allowed to sit for a week to hopefully ensure that enough time has been provided. It is not always easy to ascertain that a reaction has reached equilibrium.

Because such small molar quantities of substances will be used in this experiment, it may be convenient to work with millimolar amounts rather than molar amounts. Listed below are a few useful relationships that will be employed in the sample calculation:

**number of millimoles of solute in an aliquot of solution = $(M)(V_{mL})$**

**number of moles = $n$ = (grams)/(MM)**

**number of millimoles = mmoles = (grams)/(MM/1000) = (grams)(mMM)**

**grams = (mmoles)/(mMM) = $(n)$/(MM)**

Of course, these units are merely used as a convenience. If you prefer to use the usual molar convention, that is still equally valid.

# Sample Calculations of $K_{eq}$ for the Ethyl Acetate Hydrolysis Equilibrium

Provided below are sample calculations for the calculation of the value of $K_{eq}$ for the hydrolysis of ethyl acetate from the laboratory data. The data contained herein are typical of those you will obtain. However, the data obtained in the laboratory will differ somewhat for each student each semester. The following should be used as a guide.

***The reasoning involved in all steps in the following calculations should be thoroughly understood by each student partner in the lab pair.***

This experiment will consist of eight vials set up as described below.

| TABLE 5.2  Volumes of Each Reagent Added to the Vials | | | | | |
|---|---|---|---|---|---|
| | **3 M HCl** | **Ethyl Acetate** | **Water** | **Ethyl Alcohol** | **Acetic Acid** |
| Density of Pure Compound → | 1.055 g/mL | 0.894 g/mL | 1.00 g/mL | 0.791 g/mL | 1.05 g/mL |
| Vials 1 & 2 | 5 mL | 0 | 0 | 0 | 0 |
| Vials 3 & 4 | 5 mL | 3 mL | 2 mL | 0 | 0 |
| Vials 5 & 6 | 5 mL | 0 | 0 | 4 mL | 1 mL |
| Vials 7 & 8 | 5 mL | 4 mL | 0 | 1 mL | 0 |

Once each vial is filled with reagents as indicated, it should be labeled, capped tightly, shaken thoroughly, allowed to sit for at least 20 minutes, and then shaken one more time before storage. All eight vials will be placed into a large beaker and stored in a lab cabinet for one week. Students, their TAs, or the lab coordinator should shake the vials at least once each week until the titration laboratory period.

At the titration lab period, the vials will be titrated with NaOH (titration is simply the double replacement reaction between a strong acid and a strong base in stoichiometric amounts) to determine how much acetic acid is present. Each vial of the pair (1 & 2, 3 & 4, 5 & 6, 7 & 8) will be titrated separately, and then the average titration volume will be used to continue with calculations designed to determine the value of $K_{eq}$. Sample data are tabulated below.

**TABLE 5.3**

| Vial Number | 1 | 2 | 3 | 4 | 5 | 6 | 7 | 8 |
|---|---|---|---|---|---|---|---|---|
| mL of 1.24 M NaOH | 12.73 | 12.77 | 30.80 | 31.00 | 20.20 | 20.30 | 29.56 | 29.64 |
| Average mL NaOH | 12.75 mL | | 30.90 mL | | 20.25 mL | | 29.60 mL | |

Additional useful information includes the molar masses and millimolar masses

**TABLE 5.4**

| | EtOAc | H$_2$O | EtOH | HOAc |
|---|---|---|---|---|
| Molar Mass | 88 | 18 | 46 | 60 |
| milliMolar | 0.088 | 0.018 | 0.046 | 0.060 |

Finally, the student should recognize that Vials 1 & 2 do not contain any chemical other than the 3 M HCl. This experimental situation is called a "blank," and blanks are often analyzed as a quality control/quality assurance (QC/QA) parameter. (The 3 M HCl is an aqueous solution consisting of a small amount of HCl dissolved in water.) In this experiment, because water is actually participating as a reactant (and not just serving as an aqueous dissolving medium), information about the amount of water present in each vial can also be obtained from this titration.

## Information from the Titration of the "Blanks" (Vials 1 & 2)

mmoles of HCl in 5 mL of HCl = mmoles NaOH used in titration (***Why?***)

mmoles of HCl in 5 mL of HCl = $(C_b)(V_b)$

mmoles of HCl in 5 mL of HCl = (1.24 M)(12.75 mL)

**mmoles of HCl in 5 mL of HCl = 15.81 mmoles**

Therefore 15.81 mmoles of HCl were added (in the 5 mL of HCl) to each of the eight vials.

grams of HCl added to each of the eight vials = (mmoles)(mMM)

grams of HCl added to each of the eight vials = (15.81 mmoles)(0.03645 g/mmole)

**grams of HCl added to each of the eight vials = 0.576 g**

total mass of 3 M HCl solution = (density)(volume)

total mass of 3 M HCl solution = (1.055 g/mL)(5 mL)

**total mass of 3 M HCl solution = 5.275 g 3 M HCl**

mass of water in the 5-mL aliquot of HCl = 5.275 g solution – 0.576 g HCl

**mass of water in the 5-mL aliquot of HCl = 4.699 g**

moles of water in 5-mL HCl solution = (mass)/(MM)

moles of water in 5-mL HCl solution = (4.699 g)/(18 g/mole)

**moles of water in 5-mL HCl solution = 0.261 moles (= 261 mmoles)**

Therefore *every single one of the eight vials* contains 4.699 g, or 0.261 moles water present because of the 3 M HCl.

## $K_{eq}$ from Titration Data on Vials 3 & 4 at Equilibrium

Reaction in these vials progresses toward the production of alcohol and acetic acid. (How do we know this for sure?)

total mmoles of acid (HCl and HOAC) = mmoles of NaOH used in titration

total mmoles of acid (HCl and HOAC) = (1.24 M)(30.90 mL)

**total mmoles of acid (HCl and HOAC) = 38.32 mmoles**

mmoles HOAc = total mmoles of acid – mmoles of HCl added

mmoles HOAc = 38.32 mmoles total acid – 15.81 mmoles HCl

**mmoles HOAc = 22.51 mmoles HOAc produced in the reaction**

Therefore **at equilibrium** in Vials 3 & 4:

**mmoles HOAc present = 22.51 mmoles** = mmoles EtOAc reacting

and also mmoles HOAc present = 22.51 mmoles = mmoles water reacting *(Why?)*

**mmoles EtOH present = mmoles HOAc present = 22.51 mmoles** *(Why?)*

mmoles EtOAc present = mmoles present originally – mmoles reacting

mmoles EtOAc present = ([3 mL][0.894 g/mL])/(0.088 g/mmole) – 22.51

mmoles EtOAc present = 30.48 – 22.51 mmoles

**mmoles EtOAc present = 7.97 mmoles**

mmoles water present = total mmoles of water added – mmoles of water reacting

mmoles water present = mmoles from HCl solution + mmoles added – mmoles reacting

mmoles water present = 261 + ([2 mL][1.00 g/mL])/(0.018 g/mmole) – 22.51 mmoles

**mmoles water present = 349.60 mmoles**

The total volume is 10 mL, and molar concentration can be expressed as mmoles/mL, so

$$K_{eq} = \frac{(\text{mmoles EtOH}/10)(\text{mmoles HOAc}/10)}{(\text{mmoles EtOAc}/10)(\text{mmoles water}/10)}$$

$$K_{eq} = \frac{(22.51)(22.51)}{(7.97)(349.60)}$$

$$K_{eq} = 0.18$$

*N.B.,* **The volume cancels** *for this reaction.* **The volume will cancel in the MAE if the sum of the coefficients for reactants equals the sum of the coefficients for products in the balanced chemical equation.**

## Calculation of $K_{eq}$ from the Titration Data on Vials 5 & 6

The procedure here is similar to that with Vials 3 & 4 except that here the reaction proceeds toward the formation of ethyl acetate. (How did we decide this for certain?)

total mmoles of acid (HCl and HOAC) = mmoles of NaOH used in titration

total mmoles of acid (HCl and HOAC) = (1.24 M)(20.25 mL)

**total mmoles of acid (HCl and HOAC) = 25.11 mmoles** *at equilibrium*

mmoles HOAc = total mmoles of acid – mmoles of HCl added

mmoles HOAc = 25.11 mmoles total acid – 15.81 mmoles HCl

**mmoles HOAc = 9.30 mmoles HOAc** *at equilibrium*

mmoles HOAc reacting = mmoles HOAc originally – mmoles HOAc at equilibrium

mmoles HOAc reacting = ([1mL][1.05 g/mL])/(0.060 g/mmole) – 9.30

mmoles HOAc reacting = 17.50 – 9.30

**mmoles HOAc reacting = 8.20 mmoles**

mmoles EtOH reacting = number of mmoles HOAc reacting *(Why?)*

**mmoles EtOH reacting = 8.20 mmoles**

At equilibrium in Vials 5 & 6:

**mmoles HOAc = 9.30 mmoles**

mmoles EtOH = mmoles originally – mmoles reacting

mmoles EtOH = ([4 mL][0.791 g/mL])/(0.046 g/mmole) – 8.20

mmoles EtOH = 68.78 – 8.20

**mmoles EtOH = 60.58 mmoles**

mmoles EtOAc = mmoles HOAc reacting *(Why?)*

**mmoles EtOAc = 8.20 mmoles**

mmoles water = mmoles added (from HCl solution) + mmoles formed

mmoles water = 261 mmoles + 8.20 mmoles

**mmoles water = 269.20 mmoles**

$$K_{eq} = \frac{(60.58)(9.30)}{(8.2)(269.20)}$$

$K_{eq} = 0.25$

# Calculation of $K_{eq}$ from the Titration Data from Vials 7 & 8

Here the reaction proceeds toward the formation of acetic acid and ethanol. (How do we know this for sure?)

total mmoles of acid (HCl and HOAC) = mmoles of NaOH used in titration

total mmoles of acid (HCl and HOAC) = (1.24 M)(29.60 mL)

**total mmoles of acid (HCl and HOAC) = 36.70 mmoles *at equilibrium***

mmoles HOAc = total mmoles of acid – mmoles of HCl added

mmoles HOAc = 36.70 mmoles total acid – 15.81 mmoles HCl

**mmoles HOAc = 20.89 mmoles HOAc *at equilibrium***

At equilibrium:

**mmoles HOAc = 20.89 mmoles**

mmoles EtOH = mmoles originally added + mmoles formed by reaction

mmoles EtOH = ([1 mL][0.791])/(0.046 g/mmole) + 20.89

mmoles EtOH = 17.19 + 20.89 mmoles

**mmoles EtOH = 38.09 mmoles**

mmoles water = mmoles originally – mmoles reacted

mmoles water = 261 – 20.89 mmoles = **240.11 mmoles water**

mmoles EtOAc = mmoles originally – mmoles reacted

mmoles EtOAc = ((4 mL][0.894])/(0.088 g/mmole) – 20.89

mmoles EtOAc = 40.64 – 20.89 mmoles

**mmoles EtOAc = 19. 75 mmoles**

$$K_{eq} = \frac{(38.09)(20.89)}{(240.11)(19.75)}$$

$K_{eq} = 0.17$

The average $K_{eq}$ for the three different experiments = 0.20.

# PROCEDURE

1. Eight vials are to be labeled and then filled with the ***exact amounts*** of the chemicals indicated in Table 5.2 of this handout. Burettes or pipettes will be used to deliver the accurate aliquots of each reagent. There will be a separate burette or pipette for each reagent bottle. **Use *only* the pipette in the reagent bottle for that reagent or the burettes and other glassware used for liquid transfer. Do not contaminate the reagents by mixing up the burettes!**

2. Stopper the vials and shake for approximately 30 seconds. After 20 minutes, shake each vial again. Store the vials in a large beaker in the cabinet indicated by your TA. Return to lab each week, and shake each vial for approximately 30 seconds while you visit with your classmates.

At the lab period when you will actually titrate the contents of each vial:

3. Carefully empty the contents of Vial 1 into a 125-mL Erlenmeyer flask. Rinse the vial several times with approximately 5 mL of distilled water and add the rinsings to the Erlenmeyer flask. Add 2 to 3 drops of phenolphthalein indicator and titrate with standard NaOH until the delivery of one additional drop of the base causes a change in color from colorless to pink. The pink color must persist for 20 seconds with swirling. Record the concentration of the base and the volume required.

4. Repeat the procedure of the preceding paragraph for each of the seven other vials. The same Erlenmeyer flask can be used if the flask is rinsed well with distilled water between the titration of the contents of each vial.

# Calculations

1. Average the volumes of NaOH used for Vials 1 & 2, for Vials 3 & 4, for Vials 5 & 6, and for Vials 7 & 8. The HCl is a catalyst and is not consumed during the reaction. Thus, the amount of NaOH used in Vials 1 & 2 will also be used to titrate the HCl in each of the other vials.

2. Tabulate your data and do calculations similar to those in the sample calculation.

You should reproduce the table below in your lab notebook and fill it in with your data. You ***must*** show the calculation set-ups explicitly in order to receive full credit for your work!

| TABLE 5.5 | | | | | | | | |
|---|---|---|---|---|---|---|---|---|
| **Vial Number** | 1 | 2 | 3 | 4 | 5 | 6 | 7 | 8 |
| mL of 1.00M NaOH | | | | | | | | |
| Average mL NaOH | | | | | | | | |

# How Can a Chemical Equilibrium Be Shifted?

## Objectives

At the completion of the lab, you should be able to

- observe the effects of changing conditions on the equilibrium position of a reaction;
- identify factors that stress an equilibrium system and change the equilibrium position; and
- understand how to alter the equilibrium position of an untested reaction.

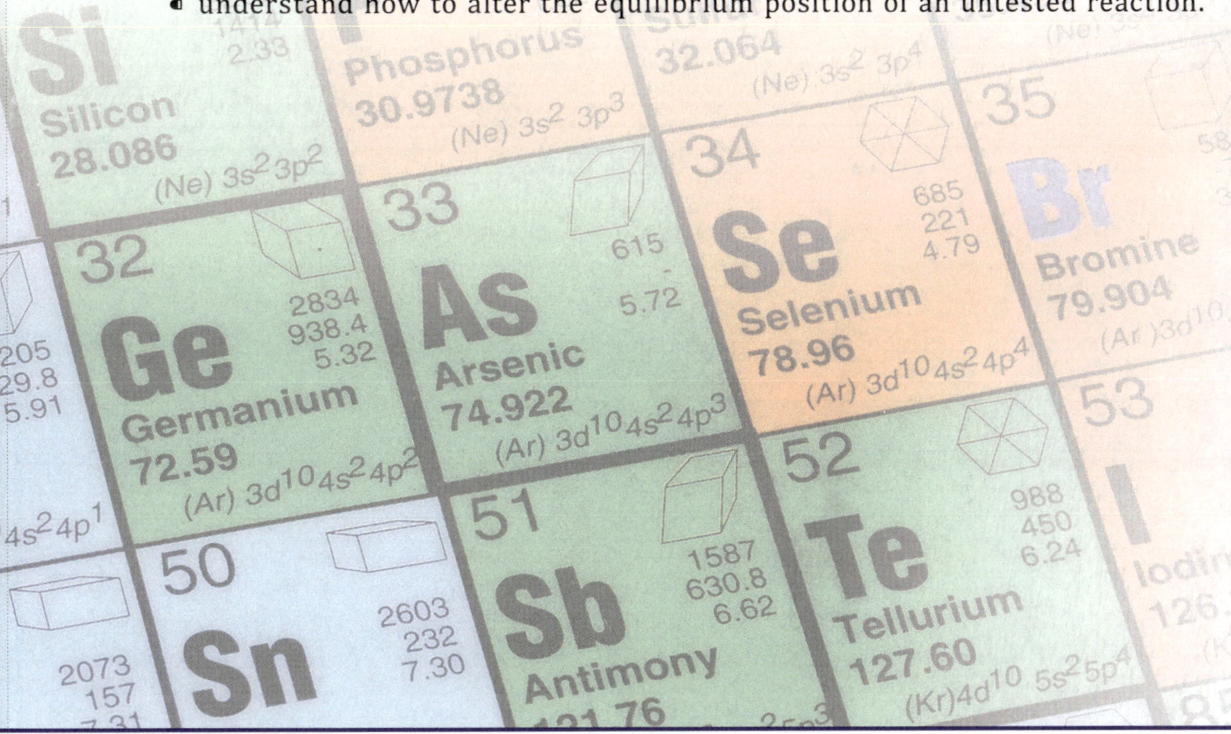

# 20-QUESTION PRE-LAB ASSIGNMENT

CRN:_________________________________________     Your Name: _______________________________________

Date Submitted: _______________________________     TA's Name: _______________________________________

*Please write your answers legibly in the space provided. Many answers will actually consist of three or four pieces of information that are clearly written on the board during the conversation. You must include all aspects of the answer to receive full credit.*

1. What are the three criteria mentioned that establish that we have an equilibrium system?

2. What are three ways to perturb an equilibrium?

3. What is the only factor (from the above three) that changes the numerical value of $K_c$?

4. How do we make a saturated solution?

5. What **must** be present in the system to prove that the solution is saturated?

6. How do we typically represent a saturated solution in CHE 112L?

**7.** How does your procedure's author represent a saturated solution?

**8.** What happens to salt added to a saturated solution?

**9.** What are the three clues that chemical change has occurred?

**10.** Adding water to our saturated salt system causes ____________________,
and the reactions shifts ____________________.

**11.** What happens when we add strong acid to our saturated solution of $Mg(OH)_2$?

**12.** An increase in temperature favors _________ processes, like _________.

**13.** A decrease in temperature favors _________ processes, like _________.

**14.** Which salts are endothermic?

**15.** Which salts are exothermic?

**16.** Which salts do we use to get metal cations into solution? Why?

**17.** What is a complex cation? What is special about them? What is our first example of a complex cation?

**18.** Tell me everything mentioned about hexane in the video,

and then tell me what happens when we add it to the $CuI_2 + I_2$ system.

**19.** Draw and label the set of sketches of test tubes that was made on the far right panel of the chalk board.

**20.** What is the azure-colored substance that appears in the last test tube sketch?

# BACKGROUND

When a chemical reaction has reached equilibrium, the concentrations of reactants and products remain constant over time. Chemical processing plants often spend much creative energy to determine how to shift equilibria so that desired products are produced in greater quantity and at a reasonable cost. How can the position of equilibrium be shifted to yield more or less products or reactants? Your goal is to identify and explain the conditions that can change the equilibrium position of a reaction.

## Procedures

You will examine the impact of changing concentration on the position of equilibrium in Parts I and II. Compare your observations with different teams, and look for patterns in results to identify factors that shift the position or equilibrium toward reactants or products. In Part III, test your understanding as you make predictions, and explain any shift in the equilibrium position of a new equilibrium.

## Teams

You will work in pairs for this lab; however, you will turn in individual reports. When developing and recording procedures, you should include enough detail that another student could reproduce your procedure based solely on your notes and directions.

<table>
<tr><td>SAFETY!<br>WASTE<br>DISPOSAL</td><td>All wastes will be collected in the Hazardous Waste container in the hood. Use a 400-mL or 600-mL beaker at your work station as satellite waste collection. At the end of the lab, transfer this waste to the Hazardous Waste container in the hood.</td></tr>
</table>

# PROCEDURE

## Part I: Observations of Equilibria: Saturated Sodium Chloride

You and your lab partner will observe and interpret the impact of changing conditions (concentration of reactants or products and addition of different reagents) on several chemical equilibrium systems and collect data. When you are done, compare your results with the other groups. In case of discrepancies in observations, repeat the procedure.

***Record all data, observations, and answers in your notebook.***

A saturated solution of sodium chloride, NaCl, is 5.4 M. What conditions will make the precipitation or NaCl($s$) as complete as possible? The reaction is

$$Na^+(aq) + Cl^-(aq) \ \leftrightarrows \ NaCl(s)$$

      colorless                  white

1. Transfer ~3 mL of saturated NaCl to each of six separate labeled (1–6) test tubes in a test tube rack. Save Tube 1 for reference.

2. Take your test tube rack to the fume hood.

<table>
<tr><td>**CAUTION!**</td><td>**Keep HCl, HNO$_3$, NaOH, and KOH off your skin and clothing. Wash any contact points with copious amounts of water.**</td></tr>
</table>

*Carefully* add concentrated HCl *one* drop at a time to Tube 2 until you have added about 5–6 drops or a change is observed. Compare the contents of this tube with your reference tube. Record the total number of drops added and your observations. Since no additional $Na^+(aq)$ ions were added, how do you account for the observations?

3. *Carefully* add concentrated nitric acid, HNO$_3$, instead of hydrochloric acid, to Tube 3. Add the same number of drops of HNO$_3$ as HCl (Step 2).

4. Compare the reactants and products from Steps 1–3. Which ion is responsible for the observed change in Tube 2? How do you know?

5. *Carefully* add a few drops of concentrated (50%) sodium hydroxide, NaOH, *one* drop at a time to Tube 4 until you have added about 5–6 drops. Record the total number of drops added and your observations. Since no additional $Cl^-(aq)$ ions were added, how do you account for the observations?

6. *Carefully* add concentrated (50%) potassium hydroxide, KOH, instead of sodium hydroxide, to Tube 5. Add the same number of drops of KOH as NaOH (Step 5). Compare the contents of this tube with the contents of the second tube. Record your observations.

7. Compare the reactants and products from Steps 5–6. Which ion is responsible for the observed change in Tube 5? How do you know?

8. Add acetone, one drop at a time, to Tube 6 until about 10 drops have been added or a change is observed. Compare the contents of this tube with the reference tube to which no reagents have been added (Tube 1). Record your observations. Since no additional $Na^+(aq)$ ions or $Cl^-(aq)$ ions were added, how do you account for the shift in the equilibrium position or the saturated sodium chloride system? (**Hint:** Acetone and water can form strong solvent–solute pairs.)

# Part II: Observations of Equilibria: Copper(II) Complex with Water and Hydroxide Ion

Copper(II) ion bonds (complexes) with water and hydroxide ion. What conditions will make the formation of a complex with hydroxide ion as complete as possible? The equilibria reactions are:

$$Cu^{2+}(aq) + 4\ H_2O(l) \leftrightharpoons [Cu(H_2O)_4]^{2+}(aq)$$

$$[Cu(H_2O)_4]^{2+}(aq) + 2\ OH^-(aq) \leftrightharpoons [Cu(H_2O)_4(OH)_2](s) + 2\ H_2O(l)$$

9. Transfer about 10 drops of 0.10 M $Cu(NO_3)_2$ to each of three small labeled (1–3) test tubes. Record your visual observations of the properties of the copper(II) complex with water.

10. Add about 10 drops of 0.10 M NaOH to each test tube. Record your visual observations or the properties of the resulting copper(II) complex with hydroxide ion. Save Tubes 1–3.

11. Stir, observe, and record your observations while you slowly add 20 drops of 1 M nitric acid, $HNO_3$, to Tube 2. What happened to the copper(II)-hydroxide complex after the addition of acid? Save Tube 2 for reference. Record your observations.

12. Stir, observe, and record your observations while you slowly add 20 drops of 1 M sodium nitrate, $NaNO_3$, to Tube 3. Record your observations.

13. Compare the reactants used and the products from Steps 11 and 12. Which ion is responsible for the observed change in Tube 2? How do you know?

14. Now reverse the order of addition of sodium hydroxide and hydrochloric acid to the copper(II) ion previously carried out (Tube 2, Steps 9 and 11). Transfer 10 drops of 0.10 M $Cu(NO_3)_2$ to a clean test tube (Tube 4). Add 20 drops of 1 M nitric acid, $HNO_3$. Stir and add about 10 drops or 0.10 M NaOH. Record your observations.

15. Compare the contents of Tubes 2 and 4 and your recorded observations. The same reactants (NaOH and $HNO_3$) were added to the copper(II) ion but in reverse order in Tubes 2 and 4. Was the observed change in Tube 2 the reverse of that observed in Tube 4? Why or why not?

# Part III: Interpreting Equilibrium Shifts Effects: Copper(II) with Iodide Ion

Use your acquired knowledge and skills to predict and interpret any change in the position of equilibrium as you alter conditions for a new reaction. The equilibrium is

$$Cu^{2+}(aq) + 4\ I^-(aq) \rightleftharpoons 2\ CuI(s) + I_2(s)$$

## Information

- $Cu^{2+}$ can react with $NH_3(aq)$ to form a deep blue ammine complex, $[Cu(NH_3)_4]^{2+}$.

- Iodine is non-polar, forming a deep purple solution when dissolved in the non-polar solvent hexane.

- The chemical equation given above for the equilibrium system omits spectator species and thus depicts the net reaction.

16. Preparation of the equilibrium system: Mix 10 mL of 0.1 M $CuSO_4$ and 10 mL of 0.1 M KI in a small Erlenmeyer flask.

17. Use your acquired knowledge and skills to explain your observations, including any shift in the equilibrium position and what happens to the copper ion upon addition of the following:

    **a.** Add 10 mL of hexane to the contents. Mix the contents thoroughly.

    **b.** Using a dropper. Add 5 M $NH_3$ to the resulting mixture contents of Step 17a. Mix the contents thoroughly and continue adding $NH_3$ until a total of 3 mL have been added.

| CAUTION! | $NH_3$ is irritating to the lungs. Perform this test in the hood. |
| --- | --- |

    **c.** Using a dropper, cautiously add 5 mL of 5 M HCl to the resulting mixture contents of Step 17b. Mix the contents thoroughly.

# DATA SHEET

Name: _______________________________________     Grade: _______________________________________

Date Experiment Performed: _______________________     Days Late: _______________________________________

CRN of Lab Section: _______________________________     Instructor's Initials: _______________________________

| General Grading Items | 20 Points |
|---|---|
| 20-Question Pre-Lab Assignment | /20 |
| Copies of Lab Pages Attached; Labeled with Name and Date, Complete Information, Readable, Data Recorded Matches Results Given in Report | |
| Waste Was Properly Disposed of and Lab Area Was Cleaned | |
| Evaluation of Student Performance Overall (Student Was on Time, Followed Safety Rules, Performed the Lab Correctly and Within the Time Allowed, Etc.) | |
| **Total** | /20 |

| Data Analysis and Interpretation | 80 Points |
|---|---|
| Part I: 5 Points/Question | /25 |
| Part II: 6 Points/Question | /24 |
| Part III: 10 Points/Observation/Explanation | /31 |
| **Total** | /80 |

# Data Analysis and Interpretation for Part I: Saturated Sodium Chloride

1. Did the addition of concentrated HCl to saturated sodium chloride shift the equilibrium toward the reactant (to the left) or product (to the right) side? How do you know?

2. HCl and $HNO_3$ are both strong acids. There is, however, a difference in the effect of these acids on the saturated sodium chloride equilibrium system (Tube 2 versus Tube 3). Explain the observed differences in these two tubes.

3. NaOH and KOH are both strong bases. There is, however, a difference in the effect of these bases on the saturated sodium chloride equilibrium system (Tube 4 versus Tube 5). Explain the observed differences in these two tubes.

4. It is a fact that when acetone is added to saturated sodium chloride, the equilibrium shifts toward the right. What observation supports this fact? What is the role of acetone in shifting the equilibrium toward the right side?

5. What reaction conditions appear to shift the position or equilibrium to favor the formation of solid sodium chloride?

# Data Analysis and Interpretation for Part II: Copper(II) Complex with Water and Hydroxide Ion

**6.** How does the addition of 1 M $HNO_3$ affect the equilibrium? Based on your observations, did the equilibrium shift toward the left or the right side? Explain.

**7.** When $HNO_3$ and $NaNO_3$ are independently added to the equilibrium system, there is a difference in observed effect (Tube 2 versus Tube 3). Identify the ion causing the equilibrium system to shift toward the left side. What is the role of this ion in shifting the equilibrium toward the left side?

**8.** Did the order of addition of nitric acid and sodium hydroxide to the equilibrium system make any difference in shifting the equilibrium (Tube 2, Steps 1 and 2 versus Tube 4, Step 5)? Why or why not?

**9.** What reaction conditions appear to shift the position of equilibrium to make the formation of a complex with hydroxide ion as complete as possible?

## Data Analysis and Interpretation for Part III

**10.** Explain your observations, including any shift in equilibrium position. Indicate what happened to the copper(II) ion as you conducted the procedure. Use chemical equations when appropriate.

# Acid–Base Titration of Buffers



## Objectives

At the completion of the lab, you should be able to

- prepare and test two acid buffer solutions; and
- determine the buffer capacity of the prepared buffers.

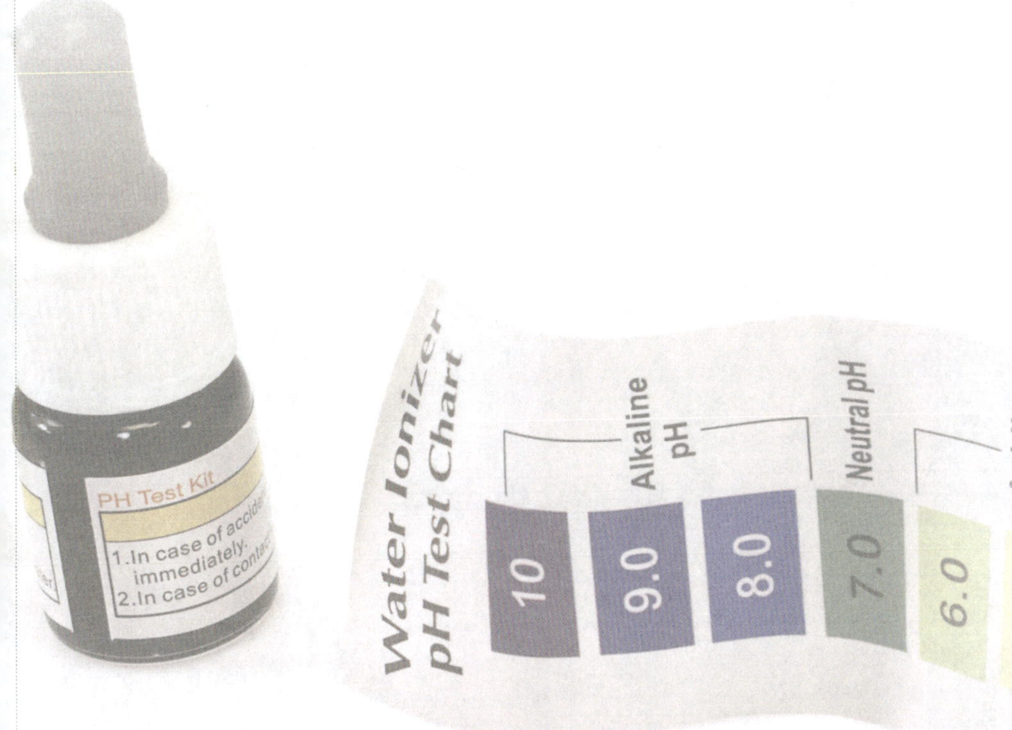

# 20-QUESTION PRE-LAB ASSIGNMENT

CRN:______________________________________     Your Name: ________________________________

Date Submitted: __________________________     TA's Name: ________________________________

*Please write your answers legibly in the space provided. Many answers will actually consist of three or four pieces of information that are clearly written on the board during the conversation. You must include all aspects of the answer to receive full credit. Happy YouTubing!*

**1.** Where are buffer systems found?

**2.** Which specific property of buffer systems are we trying to obtain this week?

**3.** What is the purpose of a buffer system?

**4.** What is the general composition of most buffer systems? What is the specific composition of the buffer system you will be studying this week?

**5.** What is the most important *equilibrium* in the buffer system?

**6.** What is the pH of pure $H_2O$? of 0.1 M HCl? of 0.1 M NaOH?

**7.** How is the buffer system equilibrium purposely different from the "usual" weak acid equilibrium system?

**8.** Set up the **I**nitial row of the **I**-C-E table for a general buffer system.

**9.** What happens to the buffer system when you add HCl (a strong acid) to it?

**10.** What happens to the buffer system when you add NaOH (a strong base) to it?

**11.** What is the definition of **buffer capacity?**

**12.** Sketch, label, and describe the typical CHE 111L titration set-up. (**N.B.,** This is the same titration set up you used to titrate HCl and HOAc in Experiment 5.)

**13.** The CHE 112L titration set-up differs slightly from CHE 111L's. What is missing?

**14.** What do we actually measure with the Logger Pro probe in a **potentiometric titration?** Use a labeled sketch of Figure 17.5 on page 600 of your textbook to illustrate the results we expect to obtain this week.

**15.** Use Figure 17.6 from page 603 of your textbook to illustrate and explain **buffer capacity.**
What is meant by the term **sigmoidal shape?**

**16.** How would one make the ideal buffer?

**17.** Write this week's buffer system in the fashion of a conjugate acid/conjugate base pair.
Show the calculation set-up, *and* then report the number of mmoles (millimoles) of acetic acid used in Buffer A this week. Finally, list the $K_a$ as well as the $pK_a$ for the acetic acid equilibrium.

**18.** The desired pH for this week's buffer is 4.0. Reproduce the calculation done in the video to determine how much conjugate base (NaOAc) needs to be added to Buffer A. Begin by first writing out the Henderson–Hasselbach equation, and then proceed to fill in the numerical values.

**19.** We need to convert the number of moles of NaOAc to the mass of NaOAc we are to add to the buffer system. Show that calculation from the video below.

**20.** What is the explicit composition of Buffer A?

# BACKGROUND

A buffer is a mixture of a weak acid and its conjugate base, or a weak base and its conjugate acid. A buffer's function is to absorb acids ($H^+$ or $H_3O^+$ ions) or bases ($OH^-$ ions) so that the pH of the system changes very, very little. By design, a buffer is an equilibrium system. For example, a buffer can be prepared with nitrous acid, $HNO_2$. The weak acid establishes an aqueous equilibrium as shown below:

$$HNO_2(aq) + H_2O(l) \leftrightarrow H_3O^+(aq) + NO_2^-(aq)$$

The equilibrium constant expression is shown below:

$$K_a = \frac{[H_3O^+][NO_2^-]}{[HNO_2]}$$

To prepare a buffer system with nitrous acid, a conjugate base is added, such as sodium nitrite ($NaNO_2$). The resulting system is a mixture of $HNO_2$ and $NO_2^-$ ions. The nitrous acid molecule will neutralize hydroxide ions, and the nitrite ion from the conjugate will neutralize hydrogen ions.

A variation of the equilibrium expression above, called the Henderson–Hasselbalch equation, is the best reference in preparing a buffer solution. For our nitrous acid/sodium nitrite buffer example, the Henderson–Hasselbalch equation is shown below.

$$pH = pK_a + \log \frac{[NO_2^-]}{[HNO_2]}$$

The pH range in which a buffer solution is effective is generally considered to be $\pm 1$ of the $pK_a$.

In this experiment, you will use the Henderson–Hasselbalch equation to determine the amount of acetic acid and sodium acetate needed to prepare two acidic buffer solutions. You will then prepare the buffers and test their buffer capacities by adding solutions of NaOH and HCl.

## Teams

You will work in pairs for this lab. However, students will turn in individual reports. The report for this lab consists of the lab report form, the data tables, and the associated graphs.

<table>
<tr><td>**SAFETY!**<br>**WASTE**<br>**DISPOSAL**</td><td>**All waste materials may be disposed in the sink.**</td></tr>
</table>

# PROCEDURE

## Part I: Preparation and Titration of Buffer Solution A

1. Use 0.149 grams of sodium acetate to prepare 100 mL of Buffer A. Record the accurate mass of sodium acetate you use and dissolve it in 100.0 mL of 0.1 M acetic acid solution.

2. Set up two burettes, burette clamps, and ring stand. Rinse and fill one burette with 0.5 M NaOH solution. Rinse and fill the second burette with 0.5 M HCl solution.

3. Use a graduated cylinder to measure out 10.0 mL of the Buffer A solution into a 250-mL beaker and add 15 mL of distilled water. Place the beaker on a magnetic stirrer, beneath the burette of NaOH, and add a stirring bar.

4. Connect a pH sensor to Channel 1 of the Vernier hand-held unit. Suspend the pH sensor in the (pH 4) Buffer A solution. Make sure that the sensor is not struck by the stirring bar.

5. The digital readout on your Vernier unit should be very close to 4.0, if not exactly equal to 4.0. Record the exact value in your data table.

6. You are now ready to titrate Buffer A. You will slowly and carefully add 0.5 M NaOH solution to the buffer solution in 0.5-mL aliquots.

    a. You already took an initial pH reading for the buffer solution before the addition of any NaOH solution in Step 5 above.

    b. Add 0.50 mL of the NaOH solution to Buffer A. When the pH stabilizes, fill in your data table with the total volume of NaOH added (0.5 mL, the current burette reading) and accurately record the pH.

    c. Continue adding the NaOH solution in 0.50-mL increments to raise the pH consistently and enter the burette reading (the total volume of NaOH added) *and* the pH. Record these data after each 0.50-mL increment has been added. Your goal is to raise the pH of the buffer by precisely two pH units.

    d. When the pH of the buffer solution is precisely two units greater than the initial reading, continue to add the NaOH solution in 0.50-mL increments until you have reached, and passed, the equivalence point of the titration.

    e. The total volume of NaOH added should be approximately 4.0 mL.

7. Dispose of the reaction mixture in the sink. Rinse the pH sensor with distilled water in preparation for the second titration of Buffer A with 0.5 M HCl.

8. Repeat Steps 3, 5, 6, and 7, using a fresh 10.0-mL sample of the Buffer A solution. For this second trial, titrate the buffer with 0.5 M HCl solution. Carefully add the HCl in 0.20-mL increments until the pH of the solution has been lowered by precisely two units. Then continue to add HCl in 0.20-mL aliquots until at least 1.4 mL total have been added. Record after each 0.20-mL aliquot addition, in the data table in your lab notebook, the total volume of HCl that has been added *and* the accurate pH reading. Prepare graphs of your data later at home.

# Part II: Preparation and Titration of Buffer Solution B

9. Use 1.490 grams of sodium acetate to prepare 100 mL of Buffer B. Record the accurate mass of sodium acetate you use and dissolve it in 100.0 mL of 1.0 M acetic acid solution. If necessary, refill the burettes containing the NaOH and HCl solutions.

10. Use a graduated cylinder to measure out 10.0 mL of the Buffer B solution and add 15 mL of distilled water. Repeat the necessary steps (5, 6, 7, and 8) to test Buffer B in a manner similar to the Part I trials. Add the 0.5 M NaOH solution in 1.0-mL aliquots to Buffer B up to a total of 20.0 mL. Record the volume of NaOH that was used to raise the pH of Buffer B by two pH units. Later at home, prepare a graph for the titration data collected using the NaOH solution.

11. Add the 0.5 M HCl solution in 1.0-mL aliquots to a second, fresh sample of 10.0 mL of Buffer B plus 15 mL of distilled water. Add the HCl solution up to a total of 16.0 mL. Record the volume of HCl that was used to lower the pH of Buffer B by two units. Later at home, prepare a graph for the titration data collected using the HCl solution.

# Data Table(s) for the Titration of a Buffer System

You will need to include Table 7.1 in your lab notebook four times for

1. Buffer A with 0.5 M NaOH;
2. Buffer A with 0.5 M HCl;
3. Buffer B with 0.5 M NaOH; and
4. Buffer B with 0.5 M HCl.

## Titrating Buffer A with 0.5 M NaOH

- You should measure the pH of the buffer system before any NaOH has been added.
- You should **add the 0.5 M NaOH to Buffer A *0.5 mL at a time.***
- You should measure the pH of the buffer system after each 0.5-mL aliquot has been added.
- You can stop NaOH addition after a total of 8.0 mL have been added.

## Titrating Buffer A with 0.5M HCl

- You should measure the pH of the buffer system before any HCl has been added.
- You should **add the 0.5 M HCl to Buffer A *0.2 mL at a time.***
- You should measure the pH of the system after each 0.2-mL aliquot has been added.
- You can stop HCl addition after a total of 4.0 mL have been added.

## Titrating Buffer B with 0.5M NaOH

- You should measure the pH of the buffer system before any NaOH has been added.

- You should **add the 0.5 M NaOH to Buffer B** for *totals of 1, 2, 3, 4, 5, 6, 10, 15, and 20 mL.*

- You should measure the pH of the system after each aliquot has been added.

- You can stop NaOH addition after 20.0 mL have been added.

## Titrating Buffer B with 0.5M HCl

- You should measure the pH of the buffer system before any HCl has been added.

- You should **add the 0.5 M HCl to Buffer B** for *totals of 1, 2, 3, 4, 5, 6, 8, 12, and 16 mL.*

- You should measure the pH of the system after each aliquot has been added.

- You can stop HCl addition after 20.0 mL have been added.

You can plot these data by hand on graph paper, *or* you can use Logger Pro (or Excel, or any other spreadsheet program).

| TABLE 7.1 | |
| --- | --- |
| **Volume of Acid (or base) Added** | **pH of the Buffer System** |
| 0.00 mL | |
| | |
| | |
| | |

*You can include as many lines in this table as you need.

# DATA SHEET

Name: _______________________________    Grade: _______________________________

Date Experiment Performed: _______________    Days Late: _______________________

CRN of Lab Section: _____________________    Instructor's Initials: ______________________

| General Grading Items | 20 Points |
|---|---|
| 20-Question Pre-Lab Assignment | /20 |
| Carbon Copies of Lab Pages **with the Data Tables;** Labeled with Name and Date, Complete Information, Readable, Data Recorded Matches Results Given in Report | |
| Waste Was Properly Disposed of and Lab Area Was Cleaned Thoroughly | |
| | |
| **Total** | **/20** |

| Data Analysis and Interpretation | 80 Points |
|---|---|
| The Four Titration Curves (10 points each) | /40 |
| Using the Data from the Titration Curves to Fill in the Data Table | /10 |
| Question 1 | /10 |
| Question 2 | /5 |
| Question 3 | /15 |
| **Total** | **/80** |

# Graphs

Plot your four graphs on the front sides of four separate sheets of paper. You are welcome to plot your graphs by hand or with a spreadsheet program (Logger Pro is the recommended spreadsheet program, but Excel or dBase are also acceptable). Attach your four graphs to the end of your lab report. Be sure that each graph is properly identified by title (e.g., pH vs. Volume of NaOH Added) and has the axes labeled. You should draw a smooth, best fit "titration curve" through your data points, as illustrated in Figures 19.7 and 19.8 on pages 855 and 859 in your textbook. Your plotted graphs can then be used to fill in the bottom of Table 7.2.

| TABLE 7.2   Data Table | Buffer A | Buffer B |
|---|---|---|
| Mass of $NaC_2H_3O_2$ Used to Prepare Buffer (grams) | | |
| Volume of Buffer Prepared (mL) | 100.0 | 100.0 |
| Molar Concentration of $HC_2H_3O_2$ in Buffer (M) | 0.1 | 1.0 |
| Initial pH of Buffer | | |
| Volume of 0.5 M NaOH to Raise pH by Two Units (mL) | | |
| Volume of 0.5 M HCl to Lower pH by Two Units (mL) | | |
| Volume of 0.5 M NaOH at Equivalence Point (mL) | | |

# Data Analysis

1. Write reaction equations to explain how your acetic acid–acetate buffer reacts with an acid and reacts with a base.

**2.** Buffer capacity has a rather loose definition, yet it is an important property of buffers. A commonly seen definition of buffer capacity is: "The amount of $H^+$ or $OH^-$ that can be neutralized before the pH changes to a significant degree." Use your data to determine the buffer capacity of Buffer A and Buffer B. (Graphically, we can identify buffer capacity by the sudden change to a very steep slope.)

**3.** Say, for example, that you had prepared a Buffer C, in which you mixed 8.203 g of sodium acetate, $NaC_2H_3O_2$, with 100.0 mL of 1.0 M acetic acid.

   **a.** What would be the initial pH of Buffer C?

   **b.** If you add 5.0 mL of 0.5 M NaOH solution to 20.0 mL each of Buffer B and Buffer C, which buffer's pH would change less? Explain.

# Qualitative Analysis
## Parts I, II, and III

# 20-QUESTION PRE-LAB ASSIGNMENT

CRN:_________________________________     Your Name: _________________________________

Date Submitted: _________________________     TA's Name: _________________________________

## Experiment 8: Qualitative Analysis: Parts I and II ———

*Please write your answers legibly in the space provided. Many answers will actually consist of several pieces of information that are clearly written on the board during the conversation. You must include all aspects of the answer to receive full credit.*

**1.** What is the **one** question qualitative analysis attempts to answer?

**2.** In that **one** question, what is **"it"**?

**3.** What must we do in order to obtain the "answer" to that **one** question?

**4.** What are we doing this week in lab? (List the parts and a "title" for each.)

**5.** What are we doing next week in lab? (List the part and a "title" for it.)

**6.** Describe what is meant by "dropwise addition of the precipitating reagent."

**7.** Take the left-to-right flow chart drawn on the board for $Ag^+$, $Hg^{2+}$, and $Pb^{2+}$ and draw it top-to-bottom in the space provided below or on a separate sheet of paper.

**8.** Sketch a map view to show what must be placed in the centrifuge across from your sample.

**9.** What does "supernatant" mean?

**10.** Which one component of the AgCl, $HgCl_2$, and $PbCl_2$ precipitate dissolves in hot water?

**11.** What did not dissolve in the hot water and is therefore still in the precipitate?

**12.** Describe what you **must** record when you run your confirmatory tests this week for $Al^{3+}$, $Cu^{2+}$, and $Zn^{2+}$.

**13.** What are the precipitating reagents **you** will use?

**14.** Uh-oh! Just realized that Dr. May wrote the incorrect chemical formula for sodium carbonate on the chalkboard! Please write the correct chemical formula for sodium carbonate.

**15.** Describe the procedure(s) you will use to test each individual cation with each precipitating reagent in lab this week.

**16.** What is the equivalent relationship between **mL** and **drops** of liquid?

**17.** Will every precipitating reagent cause a precipitate of each metal cation to form? If so, what should we do for each precipitate we observe?

**18.** Why would a precipitate that forms with a particular precipitating reagent end up dissolving at a later point in time (with the addition of an excess of that same precipitating reagent)?

**19.** What happens to a precipitate containing $AgCl$, $PbCl_2$, and $HgCl_2$ when we add excess HCl to it? Please answer very specifically.

**20.** In the space below or on a separate sheet of paper, draw the top-to-bottom flow chart from the video that shows what you will do in lab.

# PRE-LAB ASSIGNMENT

CRN:_______________________________________     Your Name: _______________________________________

Date Submitted: _______________________________     TA's Name: _______________________________________

## Experiment 9: Qualitative Analysis: Part III

*Please write your answers legibly in the space provided. This pre-lab activity has less to do with information contained in the pre-lab video and more to do with the actual chemistry of, and the chemical aspects of, qualitative analysis that we would like you to understand. You are encouraged to seek answers from any resource—your own empirical observations, the lab handout (procedure) for qualitative analysis, your textbook, other texts, people knowledgeable about qualitative analysis, etc.*

1. Fill in Table 9.1 with the information you determined empirically through the drop-wise addition of the three precipitating reagents to solutions of $Al(NO_3)_3$, $Cu(NO_3)_2$, and $Zn(NO_3)_2$ during lab. *Above the dotted line,* please write in **the *chemical formula* of the precipitate;** and *below the dotted line,* write in **the *chemical formula* of the precipitate** *or* the ***chemical formula* of the complex ion** that is present after the addition of the suggested amounts of each reagent.

| TABLE 9.1 | | | |
|---|---|---|---|
| | **Separation Reagents** | | |
| **Metal Ion** | **1 M OH$^-$** | **1 M NH$_3$** | **1 M CO$_3{}^{2-}$** |
| Al$^{3+}$ | | | |
| Cu$^{2+}$ | | | |
| Zn$^{2+}$ | | | |

2. What are *two* specific advantages to using nitrate salts to get the metal cations into solution here?

**3.** Write the three chemical equations for the three equilibria that describe the formation of the three precipitates you listed in the table in Question 1.

**4.** Interestingly, we find that the exact same precipitate forms for a particular metal cation, regardless whether we use $NaOH(aq)$, $NH_3(aq)$, or $Na_2CO_3(aq)$, as the precipitating reagent. And we name the resulting salt to reflect the anion that is present.

That anion is ________ (chemical formulation and charge) or ________ (name).

**5.** How can we explain the formation of that particular precipitate? Support your answer with chemical equation(s) showing why this anion is free in the solution.

**6.** Obtain the $K_{sp}$ values (**s**olubility **p**roduct **K**onstant) for aluminum hydroxide, copper(II) hydroxide, and zinc hydroxide, and write them below.

**7.** Which of these three hydroxides is most soluble? Least soluble? How do you know?

**8.** The $K_{sp}$ equilibrium equation is **always** written for the **dissociation** of the very slightly soluble salt. Write out the dissociation equations for aluminum hydroxide, copper(II) hydroxide, and zinc hydroxide.

**9.** Obtain the formation constants, $K_f$, for the four soluble complex ions that you observed and reported in the table in Question 1.

**10.** The $K_f$ equilibrium equation is **always** written for the **formation** of the complex ion (see examples of these equations on textbook pages 877–879). In the space below, write the formation equilibrium equation for the four soluble complex ions you included above.

**11.** Which complex ion is the most soluble? The least soluble? How do you know?

**12.** Let's say that you have an aqueous solution with the aluminum tetrahydroxide complex ion dissolved in it. The equilibrium that describes the formation of that complex ion is

$$Al(OH)_3(s) + OH^-(aq) \rightarrow Al(OH)_4^-(aq)$$

What could you do in lab to shift this equilibrium so that solid aluminum hydroxide would precipitate back out of solution? (In which direction is this shift?)

**13.** In the space below, completely describe the appearance of the positive results that you observed for the confirmatory tests you ran in lab that proved the presence of $Al^{3+}$, $Cu^{2+}$, and $Zn^{2+}$. Identify the chemical formula of the chemical substance that *is* the positive confirmatory test in each case.

**14.** In the space below or on a separate sheet of paper, write out the flow diagram that you plan to use in lab when test your unknown solution in Qualitative Analysis: Part III.

# BACKGROUND

## Procedures

This lab is designed to span two periods. This week you will complete Parts I and II, leaving Part III for the following lab meeting.

In both the confirmatory tests and separation schemes section, is it important to carefully follow directions. For example, when adding a reagent "dropwise," one drop is added; the solution is mixed and any reactions or change is noted. Then the second drop is added, etc.

Note that 20 drops is approximately 1 mL. You will need to carefully observe the results after each drop. Do not add another drop too quickly.

## Teams

You will work in pairs for Parts I and II of this lab. However, each student will analyze an individual unknown (Part III) and turn in an individual report. The report includes the Data Sheet and your lab pages. Please write neatly as the lab pages will be a large part of your grade. When developing and recording procedures, you should include enough detail that another student could reproduce your procedure based solely on your notes and directions.

<table>
<tr><td>**SAFETY!**<br>**WASTE**<br>**DISPOSAL**</td><td>**All wastes will be collected in the**<br>**Hazardous Waste container in the hood.**</td></tr>
</table>

## Introduction

Qualitative analysis involves the identification of chemical composition of unknown substances. In contrast, quantitative analysis involves a precise determination of the amount of a given substance in some sample. Say you have a sample of water and you want to know the impurities. This analysis is qualitative. Now you would like to know the concentration of these impurities, which would be a quantitative. The branch of chemistry that deals with these types of analyses is known as analytical chemistry.

In this experiment, you will determine which of several ions are present in your samples. To do this, you will not use the "traditional" qualitative analysis scheme. Instead, you will devise your own scheme for separation of your ions. Therefore, each student in your lab section may have a slightly different scheme. The only criterion is whether or not your scheme works to separate ions so that you can identify them. If so, then it is correct. It is perfectly acceptable to consult solubility tables, lists of $K_{sp}$ values, or other books on qualitative analysis to help in the preparation of your separation scheme.

In general, different ions can be compiled into groups, depending on their reaction toward a certain reagent. It is a divide and conquer approach. For example, adding a sodium hydroxide solution to some metal ions will cause no reaction in some cases (e.g., $K^+$), will yield only a precipitate in other cases (e.g., $Fe^{3+}$), and will give a precipitate that then dissolves upon the addition of more sodium hydroxide in yet other cases (e.g., $Zn^{2+}$) due to a soluble complex forming after excess $OH^-$ ions have been added. Thus, the $Fe^{3+}$ ion can be separated from $K^+$ and $Zn^{2+}$ ions by adding excess sodium hydroxide, centrifuging the container to help settle the precipitate and decanting off the solution containing the $K^+$ and $Zn^{2+}$ ions.

Summarizing this approach, two things can happen when a reagent is added to a mixture of ions. Some ions stay in solution and some precipitate. Centrifuging the test tube concentrates the solid particles in the bottom of the test tube. Then, the ions are separated by decanting (pouring) the liquid into another test tube leaving the precipitate in the original test tube. Another reagent can be added to one of the groups. Again solution or precipitation results. This can be continued until all ions are individually separated. Then a confirmatory test can be performed.

A confirmatory test occurs when a specific reagent is added to a specific ion to get a specific reaction which can be visually observed. A confirmatory test **should always** be carried out on an ion that you think has been separated from the other ions included in the analysis. Only then can you be sure you have (or don't have) that ion. Other ions may interfere with the results of the confirmatory test, so the confirmatory test is carried out **only** after the separation process has been completed.

In this experiment you will learn confirmatory tests for all of the possible ions you will deal with. You will also practice separating ions and devise a scheme for separating your ions. The only ions you will attempt to separate and identify in this experiment will be $Al^{3+}$, $Cu^{2+}$, and $Zn^{2+}$. You must prepare your own table for recording the observed result of each step you take in the confirmatory and separation processes. Be sure to completely and accurately describe everything that occurs in each solution as each reagent is added. At first your time **in** the laboratory should be spent carrying out reactions and carefully observing what happens, and your time **outside** the laboratory should be spent in trying to devise separation schemes. However, once you have a preliminary idea, it must be tested in the lab. You must be prepared to **think** in the laboratory and adjust your separation schemes as required.

## Separation Methods

The separations of metal ions will almost always involve removing a precipitate from the solution. This can be done in one of three ways depending on how complete a separation is required and whether the precipitate or the solution is of most importance.

Method one is **decantation.** Decantation is the process of carefully pouring a solution from a container, leaving the precipitate in the bottom of the container. Usually a small amount of solution must be left in the container and care must be taken to prevent a small amount of precipitate from flowing with the solution out of the container. If it is desirable to remove all of the ions that may be in the solution, the precipitate can be washed several times with an appropriate solution (one that will not dissolve the precipitate) and the wash solutions decanted.

Method two is **filtration.** Filtration is more efficient than decantation, but it is much more time consuming, and if very small amounts of solution are involved, most of the solution may be soaked up by the filter paper, thus "losing" the ions in that solution. We will not be utilizing this method in the lab.

Method three is **centrifugation.** Centrifugation involves increasing the effective gravitational force on the test tube so as to more rapidly and completely cause the precipitate to gather on the bottom on the tube. The solution is then either quickly decanted from the tube without disturbing the precipitate or withdrawn by means of a medicine dropper. **Note:** When using a centrifuge, the test tube *must always be counterbalanced* with an identical tube filled to the same level with water and placed in the position opposite the original test tube. One minute of centrifuge time is usually sufficient to settle a precipitate. Your instructor will know if a centrifuge is not counterbalanced and has the option to **deduct points due to poor technique.**

## Concepts

- Net ionic equations
- Chemical equilibria

## Chemicals

- 0.1 M solutions of copper(II) nitrate, aluminum nitrate, and zinc nitrate
- sodium hydroxide solution, 1 M
- acetic acid solution, 6 M
- ammonia solution, 1 M
- nitric acid solution, 3 M
- potassium ferrocyanide solution, 10%
- sodium carbonate solution, 1.0 M
- aluminon

## Part I: Confirmatory Tests

Practice the following tests to confirm the identity of each ion. Whenever reagents are added to a test tube, be sure they are well-mixed!

### Aluminum

When $Al(OH)_3$ precipitates in the presence of the organic dye aluminon, the dye adsorbs onto the $Al(OH)_3$ resulting in a colored, gelatinous precipitate.

1. Place 5 drops of 0.1 M aluminum nitrate in a small test tube. Add 5 drops of distilled water. Add 1 M NaOH dropwise. A precipitate of $Al(OH)_3$ forms. Add more drops of NaOH until the precipitate dissolves forming the complex ion $Al(OH)_4^{-}$.

2. Add 2–3 drops of aluminon and acidify with 6 M acetic acid. (You may use pH paper to test that the solution is acidic.) Stir the solution and allow it to stand for 1–2 minutes.

3. Add 1 M $Na_2CO_3$ dropwise until the solution is basic. (How do you know?) Centrifuge the mixture.

4. A colored precipitate (called a "lake") indicates the presence of $Al^{3+}$ ion. Carefully note the color.

## Copper(II)

5. Place 1 drop of 0.1 M copper(II) nitrate in a test tube. Add 2 mL of distilled water.

6. Add 1 drop of 3 M nitric acid to the test tube. Mix well.

7. Add 1 drop of 10% potassium ferrocyanide. Record your observations.

8. Centrifuge the mixture.

## Zinc

9. Dilute 3 drops of 0.1 M zinc nitrate with 2 mL of water and add 1 drop of 3 M nitric acid.

10. Add 1 drop of 10% potassium ferrocyanide to the test tube.

11. Record your observations.

12. Centrifuge the mixture.

# Part II: Separation Schemes

## Observations

13. Put 1 mL of 0.1 M aluminum nitrate in one test tube, 1 mL of 0.1 M copper(II) nitrate in another test tube, and 1 mL of 0.1 M zinc nitrate in a third test tube.

14. To **each** test tube add 1.0 M NaOH dropwise. If you get a precipitate, continue adding the reagent to see if the precipitate will dissolve. Probably no more than 1 mL of reagent should be added in excess; if the precipitate has not dissolved by then, it will not dissolve.

15. For each reaction that occurs, write an equation to explain what happened.

16. Be sure to thoroughly rinse out each test tube following each reaction studied.

17. Repeat Steps 13–15 above using 1.0 M $NH_4OH$ instead of the NaOH.

18. Repeat Steps 13–15 above using 1.0 M $Na_2CO_3$. ($Al(OH)_3$ will be formed here rather than $Al_2(CO_3)_3$.)

## Separation Scheme

After studying the results of these reactions and comparing them with the results from Part I for zinc, outline a scheme to completely separate aluminum, copper(II), and zinc. If, in your procedure, you get a precipitate, be sure to add something to dissolve that precipitate so that you can test for the presence of the ion. Do not forget that some reagents form soluble complexes with an ion, so you might have a solution with no precipitate present, but the ion you are testing for is still tied up in a complex ion and may not be able to give its usual confirmatory test.

19. Mix 1 mL of 0.1 M zinc nitrate, 1 mL of 0.1 M copper(II) nitrate and 1 mL of 0.1 M aluminum nitrate together.

20. Try out your separation scheme. Don't forget to **confirm** the presence of each ion when you have finished the separation!

21. If your separation scheme does not work, go back to the drawing board, rethink the problem, and devise a new scheme. You may also mix any two metal ions to test a portion of the grand scheme. The goal is to devise a scheme to separate a mixture of $Al^{3+}$, $Cu^{2+}$, $Zn^{2+}$ aqueous ions.

22. Repeat this process until you do get a successful separation. You may also mix any two metal ions.

## Summary Table for the Observed Products from the Reactions Between the Metal Cations plus the Precipitating Reagents

For each metal ion, use the top portion of each box (above the dotted line) to record the result (precipitate *or* soluble complex) you observe upon the addition of several drops of each precipitating reagent, and use the bottom portion of each box (below the dotted line) to record the result (precipitate *or* soluble complex) you observe upon the addition of up to 1 mL (~20 drops) of excess reagent.

**TABLE 8.1**

| Metal Ion | Separation Reagents | | |
|---|---|---|---|
| | 1 M OH⁻ | 1 M NH₃ | 1 M CO₃²⁻ |
| $Al^{3+}$ | | | |
| $Cu^{2+}$ | | | |
| $Zn^{2+}$ | | | |

# Part III: Determination of Unknowns

You will test the separation schemes you developed in Part II by determining which ions are present in an unknown solution. Your unknown will contain one or more of the following ions: $Al^{3+}$, $Cu^{2+}$, $Zn^{2+}$. After the separations, be sure you do a confirmatory test for each ion to prove its presence (or absence).

**23.** Obtain an unknown solution from your instructor.

**24.** Separate the ions.

**25.** Determine the ions present and absent in your unknown.

# DATA SHEET

Name: ______________________________     Grade: ______________________________

Date Experiment Performed: ______________________     Days Late: ______________________________

CRN of Lab Section: ______________________     Instructor's Initials: ______________________________

| Pre-Lab Activities | 40 Points |
|---|---|
| 20-Question Pre-Lab Assignment for Parts I and II | /20 |
| Pre-Lab Assignment for Part III | /20 |
| **Total** | **/40** |

| Parts I & II: Confirmatory Testing and Separation Schemes | 80 Points |
|---|---|
| Completion of Confirmatory Testing (**all** observations reported) | /40 |
| Reasonable Separation Scheme (must work for **all** three cations) | /40 |
| **Total** | **/80** |

| Part III: Unknown Analysis | 80 Points |
|---|---|
| Observations and Descriptions of Separations and Confirmatory Tests | /35 |
| Correct Identification of Metal Ions (15 points each) | /45 |
| **Total** | **/80** |

1. Turn in a **copy** of your tables of your observations for the confirmatory tests and for the separations schemes. Be complete—turn in all observations of all separation schemes you tried whether they were successful or not.

2. Turn in a **copy** of your successful separation schemes.

3. Turn in a report of *your* unknown that **must** include the **unknown number,** a record of the reagents you added for separations (including observations made), a record of reagents added for confirmatory tests (including observations made), and **a list of ions as shown below.** *Circle* the ion(s) present in your unknown and "X" out the ion(s) absent from *your* unknown.

   $Al^{3+}$                    $Cu^{2+}$                    $Zn^{2+}$

# The Thermodynamics of the Dissolution of Borax

## Objectives

At the completion of the lab, you should be able to

- determine the solubility product of borax as a function of temperature; and
- determine the standard free energy, standard enthalpy, and standard entropy changes for the dissolution of borax in an aqueous solution.

# 20-QUESTION PRE-LAB ASSIGNMENT

CRN:_________________________________________     Your Name: _______________________________________

Date Submitted: _______________________________     TA's Name: _______________________________________

*Please write your answers legibly in the space provided. Many answers will actually consist of several pieces of information that are clearly written on the board during the conversation. You must include all aspects of the answer to receive full credit.*

**1.** True or false? There are several varieties of borax.

What is the name, chemical formula, and solubility relative to most other sodium salts for the borax we use this week in lab?

**2.** What is the goal of thermodynamics?

**3.** What are the two possible outcomes for any chemical process?

And what does it mean when a process is "spontaneous in the **reverse** direction"?

**4.** What does "Ambient Lab Temperature and Pressure" mean?

**5.** Identify by symbol and name the thermochemical quantity we used in Chapter 6 of your textbook. And identify by symbol and name the two thermodynamic quantities that we are introduced to in Chapter 18 of your textbook.

**6.** Please write the equation that relates all three of the thermodynamic quantities from Question 5. What does "naught" mean here?

**7.** What is the special word used to describe the situation where $\Delta G < 0$? What does it mean for spontaneity when $\Delta G < 0$?

**8.** What values of $\Delta H$ favor spontaneity? (So, is spontaneity favored by endothermicity or exothermicity?) And what values of $\Delta S$ favor spontaneity? (Does this translate as greater order or greater disorder favoring spontaneity?)

**9.** What is meant by the term "monovalent salt"? Is a monovalent salt endothermic or exothermic? Does that favor or disfavor spontaneity?

**10.** For any dissolution process, what value does $\Delta S$ have? Does that favor or disfavor spontaneity?

**11.** Write out the equation that relates $\Delta G°$ to $K$ in the space below.

**12.** What values of $K$, $\ln K$, and $\Delta G$ favor spontaneity?

And what values of $K$, $\ln K$, and $\Delta G$ disfavor spontaneity?

**13.** Which specific $K$ do we use this week? And what specific value (and units) for $R$?

**14.** For a monovalent salt, what is the effect on its solubility as we decrease temperature?

**15.** What *must* be present in your test tubes to guarantee that you have a saturated solution to work with?

**16.** For the molar solubility, *s*, how *must* we write this week's reaction? Write it in that format below.

**17.** Write the equation below that relates $K_{sp}$ and *s* to each other for tincal?

**18.** What is the chemical indicator that we use this week when we titrate the borohydrate anion with HCl? Why is this indicator a better choice than the indictor phenolphthalein that we have always used in titrations before? What color change are we watching for this week?

**19.** Write down the "path" you should follow as you convert from

$$mL \text{ of } HCl \rightarrow moles\ HCl \rightarrow moles\ B_4O_5(OH)_4^{2-}$$

Show the formula you should use to calculate $s$ for the 5-mL aliquots that you titrated.

**20.** First, write the specific linear equation that you will collect data in support of and then plot. Then, draw and label a general representation of the expected linear plot you should get for your data. Finally, show explicitly how you will obtain the values for $\Delta H°$ and $\Delta S°$ that you will need in order to calculate the $\Delta G°$ value for your data.

# BACKGROUND

Large deposits of borax are found in the arid region of the southwestern United States, most notably in the Mojave Desert (east central) region of California. Borax is obtained as tincal, $Na_2B_4O_5(OH)_4 \cdot 8\,H_2O$, and kernite, $Na_2B_4O_7 \cdot 4\,H_2O$, from an open pit mine near Boron, California, and as tincal from brines from Searles Lake near Trona, California. Borax, used as a washing powder for laundry formulations, is commonly sold as 20-Mule Team Borax®. Historically, borax was mined in Death Valley, California, in the late nineteenth century. To transport the borax from this harsh environment, teams of 20 mules were used to pull a heavy wagon loaded with borax (and a water wagon) across the desert and over the mountains to railroad depots for shipment to other parts of the world. Borax is used as a cleansing agent, in the manufacture of glazing paper and varnishes, and as a flux in soldering and brazing; however, its largest current use is in the manufacture of borosilicate glass.

The free energy change of a chemical process is proportional to its equilibrium constant according to the equation

$$\Delta G° = -RT \ln K$$

Where $R$, the gas constant, is $8.314 \times 10^{-3}$ kJ/mol $\cdot K$ and $T$ is the temperature in kelvins. The equilibrium constant, $K$, is expressed for the equilibrium system when the reactants and products are in their standard states. For a slightly soluble salt in an aqueous system, the precipitate and the ions in solution correspond to the standard states of the reactants and products, respectively.

The standard state equilibrium for the slightly soluble silver chromate salt is

$$Ag_2CrO_4(s) \rightleftharpoons 2\,Ag^+(aq) + CrO_4{}^{2-}(aq)$$

The solubility product, $K_{sp}$, is set equal to the product of the molar concentrations of the ions, each raised to the power of their respective coefficients in the balanced equation—this is the mass action expression for the system:

$$K_{sp} = [Ag^+]^2[CrO_4{}^{2-}]$$

And the **free energy change** for the equilibrium is

$$\Delta G° = -RT \ln K_{sp} = -RT \ln [Ag^+]^2[CrO_4{}^{2-}]$$

Additionally, the free energy change of a chemical process is a function of the **enthalpy change** and **entropy change** of the process:

$$\Delta G° = \Delta H° - T\Delta S°$$

When the two free energy expressions are set equal for a slightly soluble salt, such as silver chromate, then

$$-RT \ln K_{sp} = \Delta H° - T\Delta S°$$

Rearranging and solving for $\ln K_{sp}$,

$$\ln K_{sp} = -\frac{\Delta H°}{R}\,(1/T) + \frac{\Delta S°}{R}$$

(which is the format of the equation of a straight line, $y = mx + b$).

This equation can prove valuable in determining the thermodynamic properties of a chemical system, such as that of a slightly soluble salt. A linear relationship exists when the values of $\ln K_{sp}$ obtained at various temperatures are plotted as a function of the reciprocal temperature. The

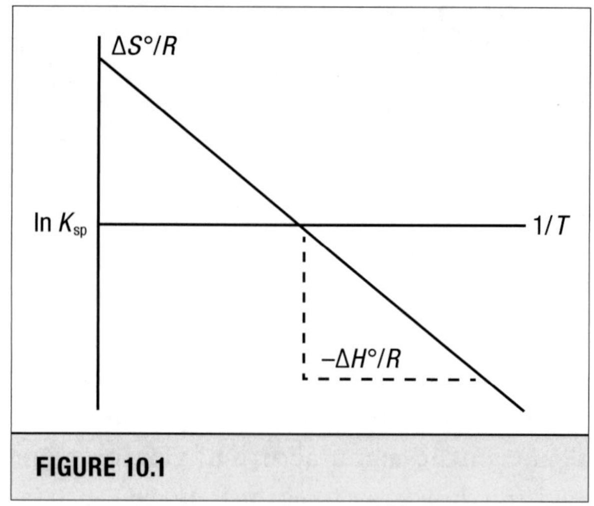


**FIGURE 10.1**

(negative) slope of the line equals $-\Delta H°/R$, and the $y$-intercept (where $x = 0$) equals $\Delta S°/R$. Since $R$ is a constant, the $\Delta H°$ and the $\Delta S°$ for the equilibrium can be readily calculated.

Since the values of $\ln K_{sp}$ may be positive or negative for slightly soluble salts and $T^{-1}$ values are always positive, the data plot of $\ln K_{sp}$ versus $1/T$ appears in the first and fourth quadrants of the Cartesian coordinate system. This is illustrated in Figure 10.1.

## The Borax System

Borax is often given the name sodium tetraborate decahydrate and the formula $Na_2B_4O_7 \cdot 10\ H_2O$. However, according to its chemical behavior, a more defining formula for borax is $Na_2B_4O_5(OH)_4 \cdot 8\ H_2O$, also called tincal.

In this experiment the thermodynamic properties, $\Delta G°$, $\Delta H°$, $\Delta S°$, are determined for the aqueous solubility of borax (tincal), $Na_2B_4O_5(OH)_4 \cdot 8\ H_2O$. Borax dissolves and disassociates in water according to the equation

$$Na_2B_4O_5(OH)_4 \cdot 8\ H_2O \rightleftharpoons Na^+(aq) + B_4O_5(OH)_4{}^{2-}(aq) + 8\ H_2O(l)$$

The mass action expression set equal to the solubility product at equilibrium, for the solubility of borax is

$$K_{sp} = [Na^+]^2[B_4O_5(OH)_4{}^{2-}]$$

The $B_4O_5(OH)_4{}^{2-}$ anion, Figure 10.2, because it is the conjugate base of the weak acid boric acid, Figure 10.3, is capable of accepting two protons from a strong acid in an aqueous solution:

$$B_4O_5(OH)_4{}^{2-}(aq) + 2\ H^+(aq) + 3\ H_2O(l) \rightleftharpoons 4\ H_3BO_3(aq)$$

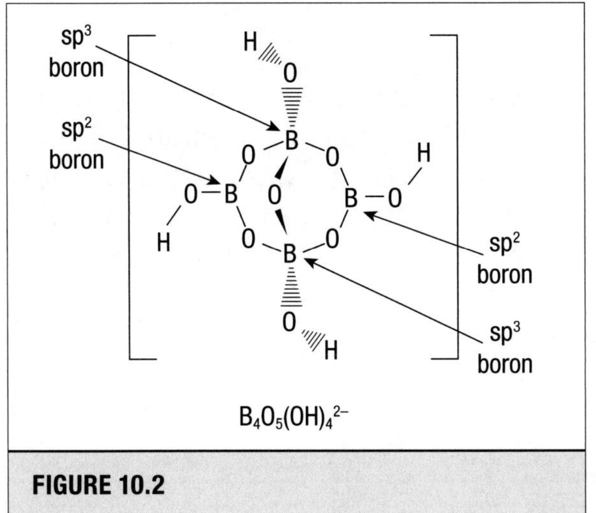


**FIGURE 10.2**

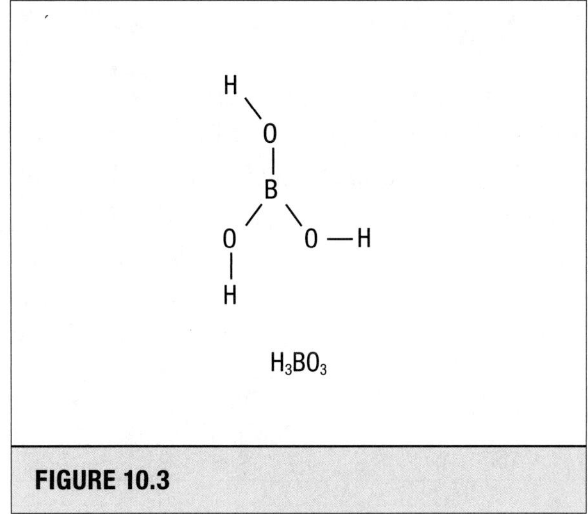


**FIGURE 10.3**

Therefore, the molar concentration of the $B_4O_5(OH)_4^{2-}$ anion in a saturated borax solution can be measured via titration of the saturated borax solution using a standardized hydrochloric acid solution as the titrant.

$$\text{mol } B_4O_5(OH)_4^{2-} = \text{volume (L) HCl} \times \frac{\text{mol HCl}}{\text{volume (L) HCl}} \times 1 \frac{\text{mol } B_4O_5(OH)_4^{2-}}{2 \text{ mol HCl}}$$

$$[B_4O_5(OH)_4^{2-}] = \frac{\text{mol } B_4O_5(OH)_4^{2-}}{\text{volume (L) sample}}$$

This analysis is also a measure of the molar solubility, $s$, of borax in water at a given temperature—according to the stoichiometry, one mole of the $B_4O_5(OH)_4^{2-}$ anion forms for every mole of borax that dissolves.

$$[B_4O_5(OH)_4^{2-}] = \text{molar solubility of borax} = s$$

Temperature changes do affect the molar solubility of most salts, and borax is no exception. For example, the solubility of borax is 2.01 g/100 mL at 0°C and is 170 g/100 mL at 100°C.

As a consequence of the titration and according to the stoichiometry of the dissolution of the borax, the molar concentration of the sodium ion in the saturated solution is twice that of the experimentally determined $B_4O_5(OH)_4^{2-}$ anion concentration.

$$[Na^+] = 2 \times [B_4O_5(OH)_4^{2-}] = 2s$$

The solubility product for borax at a measured temperature is, therefore,

$$K_{sp} = [Na^+]^2[B_4O_5(OH)_4^{2-}] = [2 \times [B_4O_5(OH)_4^{2-}]]^2[B_4O_5(OH)_4^{2-}]$$

$$= 4[B_4O_5(OH)_4^{2-}]^3 = 4[\text{molar solubility of borax}]^3 = 4s^3$$

To obtain the thermodynamic properties for the dissolution of borax, values for the molar solubility and the solubility product for borax are determined over a range of temperatures.

# PROCEDURE

The experiment is to be completed in cooperation with other chemists in the laboratory. In Part A, four warm water baths are to be set up at the beginning of the laboratory, each at a different temperature (Figure 10.4) but at a maximum of 60°C. One water bath is to be prepared by a table and then shared with everyone in lab. In conjunction with four warm water baths for Part B, about 250 mL of warm ($\cong$ 55°C) double distilled water is to be prepared for each pair of partners, too. Your table should mutually decide how you plan assemble and maintain the water bath temperature during the entire period.

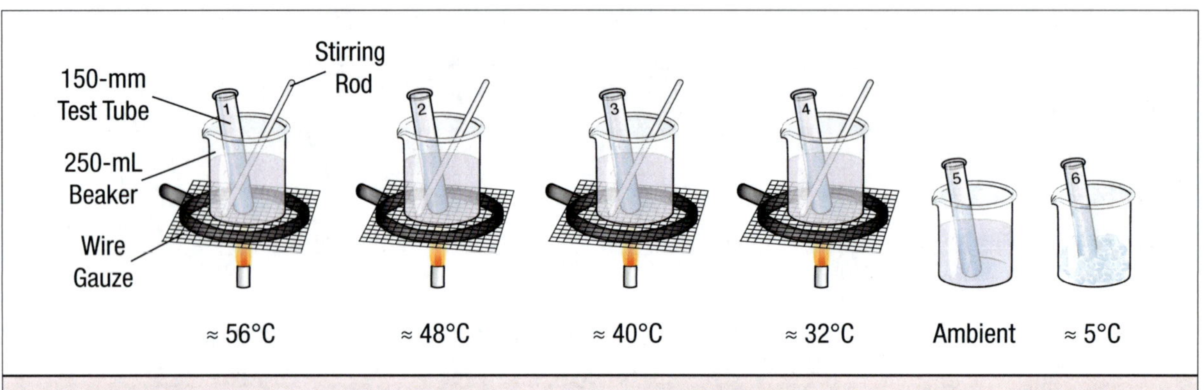


**FIGURE 10.4   Six (different temperature) Baths for Preparing Saturated Borax Solutions (Part A.3)**

# Part A: Preparation of Borax Solutions

### 1. Calibrate Test Tubes

Pipette 5 mL of distilled water into six clean medium sized (~150-mm opening) test tubes. Mark the bottom of the meniscus (use marking pen or tape). Discard the water and allow the test tubes to dry. Label the test tubes 1–6, as shown in Figure 10.5. Place this set of test tubes to the side for use in Part B.

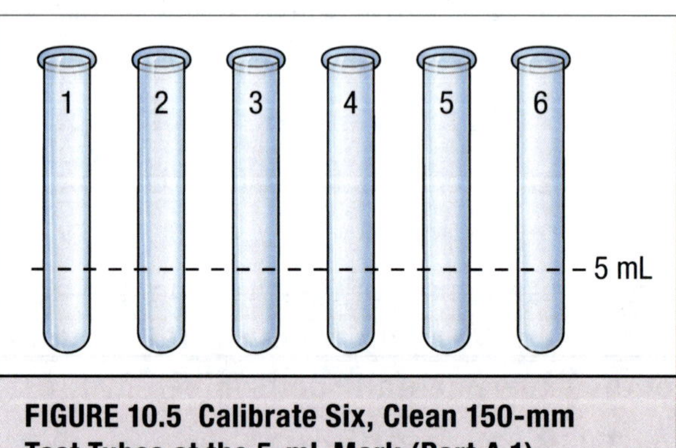


**FIGURE 10.5   Calibrate Six, Clean 150-mm Test Tubes at the 5-mL Mark (Part A.1)**

### 2. Prepare Stock Solution of Borax

Using a 125-mL or 250-mL Erlenmeyer flask, dissolve 30–35 g of borax in 100 mL of distilled water. Put a stopper in the flask, and agitate the mixture for several minutes to prepare a saturated solution.

### 3. Prepare the Test Samples of Borax

  **a.** Label a *second* set of clean test tubes 1–6.

  **b.** Thoroughly agitate the borax stock solution one more time, and half-fill the *second* set of test tubes with it. Place the test tubes in the respective baths (Figure 10.4). The temperatures of the baths need not be exactly those indicated, but the accurately measured temperature is important. The "first" bath should not exceed 60°C.

### 4. Prepare Saturated Solution of Borax

  **a.** Occasionally agitate the test tubes in the baths, assuring the formation of a saturated solution—solid borax should *always* be present in the saturated solutions—add more solid borax, if necessary. Allow the test tubes to remain in the baths 5–10 minutes.

  **b.** Allow the borax to settle until the solution (the supernatant liquid) is perfectly optically clear (this will require several minutes, be patient!) and has reached thermal equilibrium. Allow 10–15 minutes for thermal equilibrium to be established.

### 5. Prepare Borax Solutions for Analysis

Quickly, but carefully, transfer 5 mL of the supernatant liquid to the correspondingly labeled, calibrated test tubes from Part A.1. This is the first set of test tubes that you placed to the side. Record the exact temperature of the water baths. Continue to maintain water bath temperatures for others.

## Part B: Analysis of Borax Solutions

### 6. Transfer the Samples

  **a.** Set up and label a set of six clean 125-mL or 250-mL Erlenmeyer flasks. As the samples (from Part A.4) cool to room temperature in the test tubes, some borax may crystallize. Return those samples to a warm ($\leq 60$°C) water bath until all of the solid dissolves.

  **b.** After the sample is clear, transfer it to the correspondingly labeled Erlenmeyer flask, rinse the test tube with two or three 5-mL portions of warm ($\cong 55$°C) distilled water, and combine the washings with the sample.

### 7. Titrate the Samples

  **a.** Dilute each sample to about 25 mL with the warm distilled water. Add 2–3 drops of bromocresol green. Titrate each of the five samples to a yellow endpoint with the standardized 0.250 M HCl solution.

  **b.** **Waste disposal** will be specified by the lab manager and announced.

## Data Analysis

Eighteen data calculations (the molar solubility, $s$, and the $K_{sp}$ at each of the six bath temperatures) are required in order to establish the data plot for the determination of $\Delta H°$ and $\Delta S°$ for the dissolution of borax. *You must show all your calculations* on a separate sheet of lined paper! The data from the calculations can then be plotted using Logger Pro. The calculations and the analysis that are required are:

8. Calculate the molar solubility of borax at each of the measured temperatures.

9. Calculate the solubility product of borax at each of the measured temperatures.

10. Plot the natural logarithm of the solubility product versus the reciprocal temperature ($K^{-1}$) for each sample, autoscale your graph, and have Logger Pro "draw" the "best straight line" through the data points for you.

11. Obtain the slope of the linear plot, and calculate the standard enthalpy of solution for borax.

12. Obtain the $y$-intercept (at $x = 0$) of the linear plot, and calculate the standard entropy of solution for borax.

# DATA SHEET

Name: _______________________________     Grade: _______________________________

Date Experiment Performed: _______________     Days Late: _______________________________

CRN of Lab Section: _______________________     Instructor's Initials: _______________________

| General Grading Items | 20 points |
|---|---|
| Copies of the Lab Notebook Data Pages Submitted to TA | |
| Waste Was Properly Disposed of and Lab Area Was Cleaned | |
| Student Behavior: On-Time Arrival, Followed Safety Rules, Polite and Courteous | |
| 20-Question Pre-Lab Assignment | /20 |
| **Total** | /20 |

## TABLE 10.1  Preparation of Borax Solution

| Sample Number | 1 | 2 | 3 | 4 | 5 | 6 |
|---|---|---|---|---|---|---|
| Volume of Sample | 5 | 5 | 5 | 5 | 5 | 5 |
| Temperature of Sample | | | | | | |

## TABLE 10.2  Analysis of Borax Test Solutions

| Sample Number | 1 | 2 | 3 | 4 | 5 | 6 |
|---|---|---|---|---|---|---|
| Burette Reading, Final (mL) | | | | | | |
| Burette Reading, Initial (mL) | | | | | | |
| Volume of HCl Added (mL) | | | | | | |

## TABLE 10.3  Data Analysis

| Sample Number | 1 | 2 | 3 | 4 | 5 | 6 |
|---|---|---|---|---|---|---|
| Temperature (K) | | | | | | |
| $1/T$ $(K^{-1})$ | | | | | | |
| Moles of HCl Used | | | | | | |
| Moles of $B_4O_5(OH)_4^{2-}$ | | | | | | |
| $[B_4O_5(OH)_4^{2-}]$ (mol/L) | | | | | | |
| Molar Solubility of Borax | | | | | | |
| Solubility Product, $K_{sp}$ | | | | | | |
| $\ln K_{sp}$ | | | | | | |

## Questions

1. Instructor's approval of plotted data:

2. From the data plot determine

   **a.** $-\Delta H°/R$ (the slope):

   **b.** $\Delta S°/R$ (the $y$-intercept):

   **c.** $\Delta H°$ (kJ/mol):

   **d.** $\Delta S°$ (J/mol $\cdot$ $K$):

   **e.** $\Delta G°$ (kJ), at 298 K:

# Electrochemical Cells and Standard Electrode Potentials

# 20-QUESTION PRE-LAB ASSIGNMENT

CRN:_________________________________________   Your Name: _______________________________________

Date Submitted: _______________________________   TA's Name: _______________________________________

*Please write your answers legibly in the space provided. Many answers will actually consist of several pieces of information that are clearly written on the board during the conversation. You must include all aspects of the answer to receive full credit.*

**1.** What is electricity?

**2.** What are the two possible outcomes for an electrochemistry situation?

**3.** What is the classic example of a *redox* process? (We saw it in CHE 111L, too.)

**4.** What type of reaction is the one shown above? And what is sulfate anion in this process?

**5.** Write the chemical equation again, this time assigning oxidation numbers where warranted, showing the atom reduced and the oxidizing agent, and showing the atom oxidized and the reducing agent.

**6.** Why is it preferred to write the oxidation half-reaction first?

**7.** Write the two half-reactions for this example, and include the $E°$ values?

**8.** How did we predict/assess spontaneity last week? How will we do it this week?

**9.** In Table 11.1 Standard Reduction Potentials at 25°C, how do we designate "standard thermodynamic conditions"? What are those conditions?

**10.** Define "reduction." Where do we show the electrons in the reduction half-reaction?

**11.** What value(s) of $E°$ in the table mean "spontaneous reduction (as written)"?

**12.** How many electrons, $n$, are transferred in our classic example?

**13.** What does the value of $E° = -0.76$ V for the process $2e^- + Zn^{2+} \rightarrow Zn(s)$ as written in the table tell us?

**14.** Draw and label the Daniell cell below.

**15.** When does a battery stop "working"?

**16.** Why is RC Cola the "electroChemist's Cola"?

**17.** In general, which metal should be the cathode, and which metal should be the anode in a galvanic cell?

**18.** What will our ambient lab conditions be this week? Are they "standard"?

**19.** Find the Nernst equation in your text and write it below.

**20.** In the space below, show the calculation of $E^\circ_{cell}$ for the Daniell cell.

# BACKGROUND

In electrochemistry, a galvanic cell is a specially prepared system in which an oxidation-reduction reaction occurs spontaneously. This spontaneous reaction produces an easily measured electrical potential.

In this experiment, you will construct several galvanic (or voltaic) cells and measure the voltage produced these cells. A voltaic cell is constructed by using two metal electrodes and solutions of their respective salts (the electrolyte component of the cell) with known molar concentrations. You will use a voltage probe to measure the potential of a voltaic cell with copper, lead, and zinc electrodes. This measure voltage will be compared to calculated standard cell potentials.

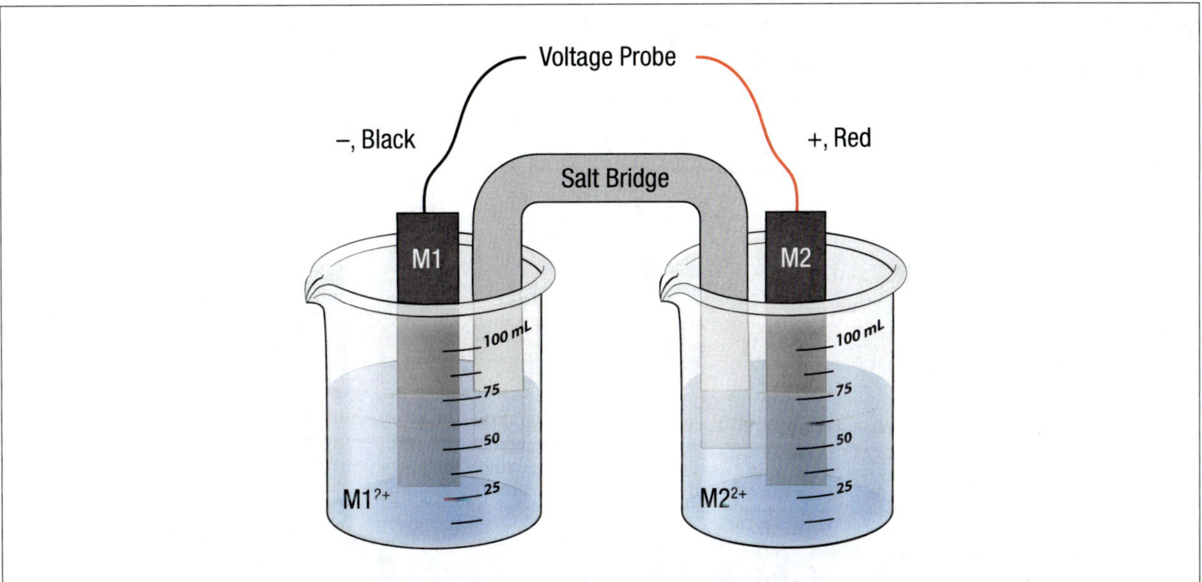


**FIGURE 11.1 Measuring Cell Potentials Using the Galvanic Cell Model** The beaker on the left contains metal one (M1) and an aqueous solution of metal one ions, and the beaker on the right contains metal two (M2) and a solution of its ions.

**TABLE 11.1  Standard Reduction Potentials at 25°C**

| Reduction Half-Reaction | | | $E°$(V) |
|---|:---:|---|:---:|
| $F_2(g) + 2e^-$ | $\rightarrow$ | $2\ F^-(aq)$ | 2.87 |
| $H_2O_2(aq) + 2\ H^+(aq) + 2e^-$ | $\rightarrow$ | $2\ H_2O(l)$ | 1.78 |
| $MnO_4^-(aq) + 8\ H^+(aq) + 5e^-$ | $\rightarrow$ | $Mn^{2+}(aq) + 4\ H_2O(l)$ | 1.51 |
| $Cl_2(g) + 2e^-$ | $\rightarrow$ | $2\ Cl^-(aq)$ | 1.36 |
| $Cr_2O_7^{2-}(aq) + 14\ H^+(aq) + 6e^-$ | $\rightarrow$ | $2\ Cr^{3+}(aq) + 7\ H_2O(l)$ | 1.33 |
| $O_2(g) + 4\ H^+(aq) + 4e^-$ | $\rightarrow$ | $2\ H_2O(l)$ | 1.23 |
| $Br_2(aq) + 2e^-$ | $\rightarrow$ | $2\ Br^-(aq)$ | 1.09 |
| $Ag^+(aq) + e^-$ | $\rightarrow$ | $Ag(s)$ | 0.80 |
| $Fe^{3+}(aq) + e^-$ | $\rightarrow$ | $Fe^{2+}(aq)$ | 0.77 |
| $O_2(g) + 2\ H^+(aq) + 2e^-$ | $\rightarrow$ | $H_2O_2(aq)$ | 0.70 |
| $I_2(s) + 2e^-$ | $\rightarrow$ | $2\ I^-(aq)$ | 0.54 |
| $O_2(g) + 2\ H_2O(l) + 4e^-$ | $\rightarrow$ | $4\ OH^-(aq)$ | 0.40 |
| $Cu^{2+}(aq) + 2e^-$ | $\rightarrow$ | $Cu(s)$ | 0.34 |
| $Sn^{4+}(aq) + 2e^-$ | $\rightarrow$ | $Sn^{2+}(aq)$ | 0.15 |
| $2\ H^+(aq) + 2e^-$ | $\rightarrow$ | $H_2(g)$ | 0 |
| $Pb^{2+}(aq) + 2e^-$ | $\rightarrow$ | $Pb(s)$ | −0.13 |
| $Ni^{2+}(aq) + 2e^-$ | $\rightarrow$ | $Ni(s)$ | −0.26 |
| $Cd^{2+}(aq) + 2e^-$ | $\rightarrow$ | $Cd(s)$ | −0.40 |
| $Fe^{2+}(aq) + 2e^-$ | $\rightarrow$ | $Fe(s)$ | −0.45 |
| $Zn^{2+}(aq) + 2e^-$ | $\rightarrow$ | $Zn(s)$ | −0.76 |
| $2\ H_2O(l) + 2e^-$ | $\rightarrow$ | $H_2(g) + 2\ OH^-(aq)$ | −0.83 |
| $Al^{3+}(aq) + 3e^-$ | $\rightarrow$ | $Al(s)$ | −1.66 |
| $Mg^{2+}(aq) + 2e^-$ | $\rightarrow$ | $Mg(s)$ | −2.37 |
| $Na^+(aq) + e^-$ | $\rightarrow$ | $Na(s)$ | −2.71 |
| $Li^+(aq) + e^-$ | $\rightarrow$ | $Li(s)$ | −3.04 |

# PROCEDURE

- Construct and measure the potential of a galvanic cell.

- Use standard reduction potentials to calculate standard cell potentials.

- Use the Nernst equation to calculate cell potentials under non standard conditions.

1. Use two 50-mL beakers and a salt bridge to construct your voltaic cell. The salt bridge is a piece of candle wicking that is saturated with $KNO_3(aq)$. Obtain 20–30 mL of 1 M $KNO_3(aq)$ in a small beaker and use a disposable pipette to wet your candle wicking.

2. Obtain one of each metal strip to act as electrodes. Polish each strip with sand paper. Place each metal strip in a 50-mL beaker containing 20 mL of a 0.1 M solution the respective metal ion. These are the half-cells.

3. Connect a voltage probe to Channel 1 of the Vernier computer interface. Connect the interface to the computer with the proper cable and start Logger Pro.

4. Connect the probe clamps to the metal strips, measure the potential of each voltaic cell, and complete Table 11.2.

5. Use Table 11.1 Standard Reduction Potentials at 25°C (posted in Blackboard (Bb)) to find $E°$ and to calculate $E°_{cell}$.

**TABLE 11.2 Template Table for Notebook**

| Electrodes | Half-Reactions | $E_{cell}$ (measured) | $E°$ | $E°_{cell}$ |
|---|---|---|---|---|
| Cu | | | | |
| Zn | | | | |
| Cu | | | | |
| Pb | | | | |
| Pb | | | | |
| Zn | | | | |

# DATA SHEET

Name: _______________________________     Grade: _______________________________

Date Experiment Performed: _______________     Days Late: _______________________

CRN of Lab Section: _______________________     Instructor's Initials: _______________

|  |  |
| --- | --- |
|  |  |
|  |  |
|  |  |

| General Grading Items | 20 Points |
| --- | --- |
| 20-Question Pre-Lab Assignment |  |
|  |  |
|  |  |
| **Total** | **/20** |

| Table of Cell Potentials | 54 Points |
| --- | --- |
| 12 Points per Constructed Cell: 6 Points for Calculation Standard Potential, 3 Cells | /18 each |
| **Total** | **/54** |

| Questions | 54 Points |
| --- | --- |
| Question 1: 6 Points | /6 |
| Question 2: 6 Points | /2 |
| Question 3: 6 Points (4 points each for 3 cells) | /12 |
| Question 4: 6 Points | /6 |
| **Total** | **/26** |

**TABLE 11.3   Table of Measured and Observed Cell Potentials**

| Electrodes | Half-Reactions | $E_{cell}$ (measured) | $E°$ | $E°_{cell}$ |
|---|---|---|---|---|
| Cu | | | | |
| Zn | | | | |
| Cu | | | | |
| Pb | | | | |
| Pb | | | | |
| Zn | | | | |

# Questions

**1.** What conditions are required for a cell to be a "standard" cell?

**2.** Are the cells you constructed standard cells?

**3.** Use the Nernst equation to calculate $E_{cell}$ for the three cells you constructed.

**4.** Explain why the measured cell potentials are essentially equivalent to the calculated cell potentials.

# Sample Calculation and Cooling Curves

Below is a cooling curve similar to the ones you will produce in lab. The red trace is for pure water, and the blue trace is an aqueous urea solution. Notice the temperature drops below the freezing point, increases sharply, and then levels off (recall temperature is constant during a phase change). The temperature drops below the freezing point due to super cooling and the abrupt increase in temperature is due to the onset of freezing. The temperature readings in the bracketed portion of each curve below were averaged in Logger Pro and are used to determine the freezing points of the respective solutions.

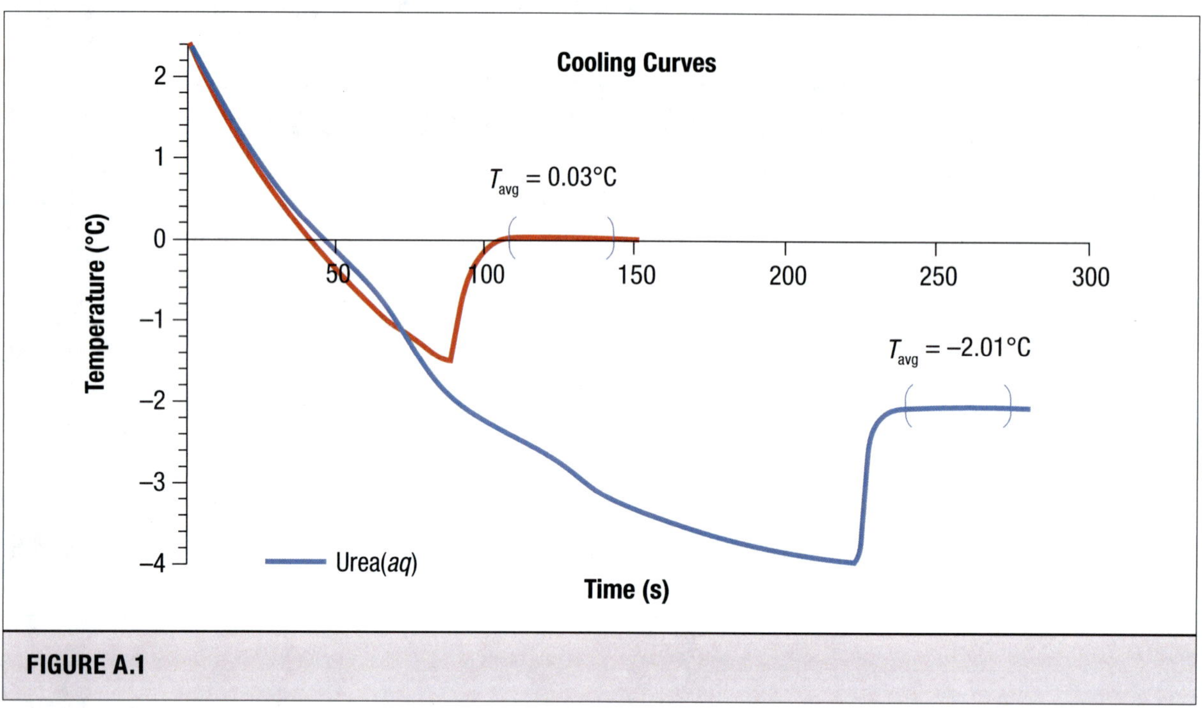


**FIGURE A.1**

# Experimental Determination of $K_f$ for Water

The freezing point depression constant for water can be determined by using the data in the cooling curve above. The freezing point of distilled water is $T_{avg} = 0.03°C$.

| TABLE A.1 | | | | | |
|---|---|---|---|---|---|
| **Trial** | **Urea Solution** | | | $\Delta T_f$/°C | $K_{f,expt}$/°C · m$^{-1}$ |
| | **Mass of Urea/g** | **Mass of Water/kg** | **Molality/m** | | |
| 1 | 0.997 | 0.015078 | 1.102 | 2.04°C | 1.85 |

The above calculated freezing point depression constant is in good agreement with the accepted value of 1.86°C/m. The value of $K_f$ **that is calculated** will then be used in Part III to determine the van't Hoff factor of an unknown electrolyte.

## Sample Calculations

molality:

$$m_{urea} = \frac{n_{urea}}{\text{mass water in kg}}$$

$$m_{urea} = 0.997 \text{ g} \times \frac{1 \text{ mol}}{60.0 \text{ g}} \times \frac{1}{0.015078 \text{ kg}}$$

$$m_{urea} = 1.102 \text{ mol} \cdot \text{kg}^{-1}$$

$\Delta T_f$:

$$\Delta T_f = \Delta T_{f,water} - \Delta T_{f,urea}$$

$$\Delta T_f = 0.03°C - (-2.01°C)$$

$$\Delta T_f = 2.04°C$$

$K_f$:

$$\Delta T_f = iK_f m$$

$$K_f = \frac{\Delta T_f}{im}$$

$$K_f = \frac{2.04°C}{1 \cdot 1.102 \text{ mol} \cdot \text{kg}^{-1}}$$

$$K_f = 1.85°C/m$$

# Colorimeter



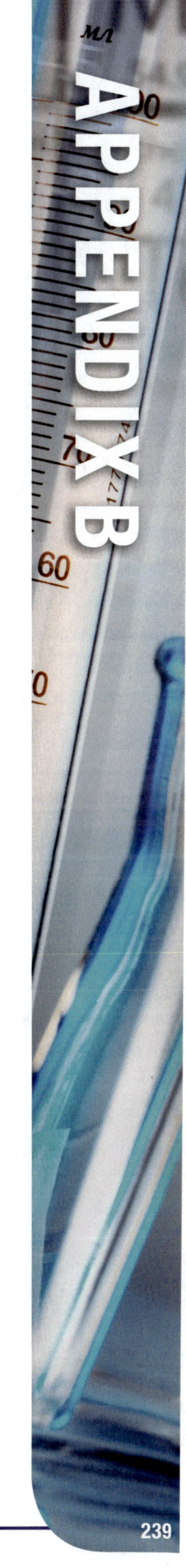

You will use a colorimeter (a side view is shown in Figure B.1) to measure the absorbance of the bleach/dye mixture. You can select one of four LED light wavelengths with the Vernier Colorimeter; violet (430 nm), blue (470 nm), green (565 nm), and red (635 nm). In this experiment the red (635 nm) will be used. This LED light will pass through the solution and strike a photocell. A higher concentration of the colored solution absorbs more light (and transmits less) than a solution of lower concentration. The colorimeter monitors the light received by the photocell as percent transmittance.

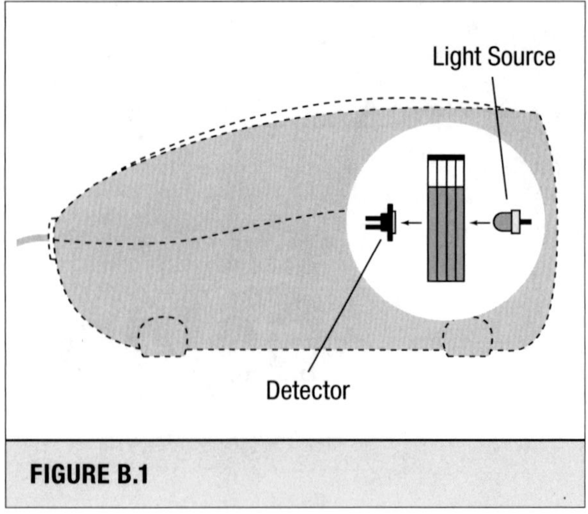


**FIGURE B.1**

# How to Use the Colorimeter

## Calibrate the Colorimeter

1. Prepare a *blank* by filling an empty cuvette ¾ full with distilled water. Place the blank in the cuvette slot of the Colorimeter and close the lid.

2. Set the wavelength on the colorimeter to the wavelength at which you will make your absorption readings and press the "CAL" button. The red LED should begin to flash. When the LED stops flashing (the absorbance should read between 0.000 or 0.001), the calibration is complete and your unit is ready to collect data.

## Absorption Measurement

3. You are now ready to collect absorbance-concentration data for the bleach/dye mixture.

   a. Make sure your cuvette is clean and dry. Prepare the reaction mixture and described in section three. Quickly fill the cuvette ¾ full and place in the sample compartment and click ▶ Collect .